D0773434

OIL
The Price of Power

THE PRICE OF POWER

Edited by
ADRIAN HAMILTON

Michael Joseph/Rainbird
in association with
Channel Four Television Company Limited

To A. Strindberg

First published in Great Britain by Michael Joseph Ltd
27 Wrights Lane, Kensington, London W8 5TZ in association
with The Rainbird Publishing Group Ltd
who designed and produced the book

British Library Cataloguing in Publication Data
Oil: the price of power.
1. Petroleum industry and trade – History
I. Hamilton, Adrian
338.2'7282'09 HD9560.5

ISBN 0-7181-2783-8

Text set by Tradespools Limited, Frome, Somerset.
Illustrations originated by Mullis Morgan Limited, London.
Printed and bound by Butler and Tanner Limited, Frome, Somerset.

Contents

Maps and Diagrams

Foreword

This is not just the story of the world's largest industry: it is a history of the modern world, as seen through the most global and political of all businesses. It shows up familiar events – including two world wars, the development of the Middle East, the boom in Texas and the crisis in Mexico – through the calculating eyes of oilmen preoccupied with the problems of profits and margins. And on the other side it shows how politicians, generals and revolutionaries have seen their fate decided by the movement of oil and the laws of supply and demand.

Among many histories of oil this book has a special topicality and relevance at a time when the oil price, on which the prosperity of so many nations depends, has once again become fragile and unpredictable. It adds an unusual insight, written by an author with a long experience of the industry, making use of revealing interviews provided for the television series *Oil*. Many of the chief actors in the oil dramas give their own accounts of events at the time. Dr Armand Hammer of Occidental describes how *he* saw his deals with Colonel Gadaffi which first broke the unity of the 'Seven Sisters'; Sheikh Yamani provides his own assessment of the role of Saudi Arabia and the future of the oil price; members of the Hunt family in Texas give their version of the family wrangles which provided the original for *Dallas*; David Rockefeller gives his own account of his grandfather who established the first great oil monopoly, Standard Oil.

This book can thus provide an authoritative account of the dramatic turning-points of oil history, as it swings between shortage and glut, between cartels and confusion; while adding vivid sidelights from living participants who reflect on its significance. It appears at a time when oil experts are having to question much of the accepted wisdom of their business and when ordinary people have become aware of how much their own daily lives and annual budgets depend on the price and behaviour of this slippery fuel. To try to understand the shocks of the future it is essential to look at the shocks of the past.

Anthony Sampson

1

'God Bless Standard Oil'

When 'Colonel' Drake started it all with his discovery of oil at Titusville in Pennsylvania in 1859, he sold the liquid as lighting fuel at $20 a barrel. Within two years the price was down to 10 cents per barrel as a speculative onslaught seized every available piece of drilling land in that part of the United States. Three years later, the wells exhausted, the price had risen to $14 per barrel. By 1865 it was back down to $4, only to veer up and down through the rest of the century until the Spindletop discovery in Texas in 1901, the greatest gusher of all time, had it down to 25 cents per barrel and below. Up again before the First World War, the price went in all directions between the wars, slumped in the post-war years, rocketed in the seventies and now, in the eighties, it is once more gyrating down the scale.

Oil has never been a predictable commodity, although it seems that every generation has to learn that anew. Nor has it always brought wealth, let alone happiness, to those who have discovered it. 'Colonel' Drake – he was never a real colonel but a former conductor on the New Haven Railroad – ended his life in poverty, the first flush of wealth drained away in a spectacular rash of new oil ventures which never paid off. Captain Lucas and Colonel Guffey (their titles were real), who pioneered Texas oil, ended with a pittance, complaining bitterly that they had been cheated out of their rights by the money men who moved in to take advantage of the Spindletop gusher.

Predictable or not, oil has been the most important resource of our time. Without it the advances of the twentieth century would not have been possible. The car, the aircraft, the tank and the space shuttle could never have happened. Communications, transport, mechanized warfare and the mobility of man have made this century what it is and they are there because oil has powered them.

It has been the struggle to organize and control this power that has made oil so particularly a political commodity and such a contentious one. 'Chaos and disorder, waste and incompetence, competition at its worst,' commented 21-year-old John Davison Rockefeller (John D. Rockefeller as history was to know him) as he surveyed the greed, the clamour and the sheer energy of the thousands of people pouring into Pennsylvania to take advantage of Drake's discovery. Yet Rockefeller's campaign to bring his order out of their chaos was to mark for ever the industry which he founded with the image of the predator and the profiteer.

The search for stability both of price and supply, has been continual and nearly always futile. Oil may be too important a commodity to be left in the hands of the market and the oil companies which have dominated it for so long but it is too mercurial a liquid ever to be subject to neat arrangements. The oil companies have tried to achieve stability through integration of the entire operation from exploration to marketing as well as through cartel arrangements among them-

Year	*Price $/barrel*	*Year*	*Price $/barrel*	*Year*	*Price $/barrel*	*Year*	*Price $/barrel*
1860	9·59	1892	·56	1924	1·43	July 1953	1·93
1861	·49	1893	·64	1925	1·68	1954	1·93
1862	1·05	1894	·84	1926	1·88	1955	1·93
1863	3·15	1895	1·36	1927	1·30	1956	1·93
1864	8·06	1896	1·18	1928	1·17	June 1957	2·08
1865	6·59	1897	·79	1929	1·27	1958	2·08
1866	3·74	1898	·91	1930	1·19	February 1959	1·90
1867	2·41	1899	1·29	1931	·65	August 1960	1·76
1868	3·63	1900	1·19	1932	·87	September 1960	1·80
1869	3·64	1901	·96	1933	·67	1961	1·80
1870	3·86	1902	·80	1934	1·00	1962	1·80
1871	4·34	1903	·94	1935	·97	1963	1·80
1872	3·64	1904	·86	1936	1.09	1964	1·80
1873	1·83	1905	·62	1937	1·18	1965	1·80
1874	1·17	1906	·73	1938	1·13	1966	1·80
1875	1·35	1907	·72	1939	1·02	1967	1·80
1876	2·56	1908	·72	1940	1·02	1968	1·80
1877	2·42	1909	·70	1941	1·14	1969	1·80
1878	1·19	1910	·61	1942	1·19	1970	1·80
1879	·86	1911	·61	1943	1·20	15 February 1971	2·18
1880	·95	1912	·74	1944	1·21	1 June 1971	2·29
1881	·86	1913	·95	1945	1·05	20 January 1972	2·48
1882	·78	1914	·81	1946	1·05	1 January 1973	2·59
1883	1·00	1915	·64	December 1946	1·20	1 April 1973	2·74
1884	·84	1916	1·10	March 1947	1·60	1 June 1973	2·90
1885	·88	1917	1·56	December 1947	2·20	1 July 1973	2·96
1886	·71	1918	1·98	July 1948	1·99	1 August 1973	3·07
1887	·67	1919	2·01	April 1949	1·84	1 October 1973	3·01
1888	·88	1920	3·07	July 1949	1·71	16 October 1973	5·12
1889	·94	1921	1·73	1950	1·71	1 November 1973	5·18
1890	·87	1922	1·61	1951	1·71	1 December 1973	5·04
1891	·67	1923	1·34	1952	1·71		

1860–1899 Pennsylvania Crude Prices
1900–1944 US Average Crude Price
1945–1973 Arabian Light Posted at Ras Tanura
From 1957 to 1972 market prices of crude oils were at a discount on posted prices.
In the period 1961 to 1970 market prices were in the range $1·3 to $1·5 barrel.

Source: British Petroleum

The course of oil prices from 1860 to 1973, showing the intense volatility in the early decades, the impact of Spindletop (1901), the East Texas discoveries (1930), and the influence of OPEC in the years before the doubling of prices in 1973.

selves, only to see new discoveries and new competitors erode the market. For their part, the consumers have found that the twin aims of security of supply and low cost are rarely compatible.

The producing countries have gnawed at the problem of controlling their own wealth, to discover that the wheel of fortune has them at the top in one decade only to turn them to the bottom the next. 'Really oil is almost a noble product,' declared the Shah of Iran, justifying a doubling of its price in December 1973. Just over five years later he was toppled from power, setting off an even more extreme price rise. Within fifteen years oil was running out of the world's ears and the price was back down, in effective terms, to where it had been before the Shah's speech.

If the United States has always dominated this trade, through its industry and its importance in the market, this is not merely because the technology, the companies and the logistics were first developed there. It is also because most of the factors that have controlled the business were originally experienced and tackled there.

Snake Oil and Greek Fire

Oil was hardly an unknown product before Drake drilled his famous well at Titusville. 'Make thee an ark,' runs Genesis 6:14, '. . . and . . . pitch it within and without with pitch'; or, more appropriately, considering the later history of the industry, Genesis 14:10 later declares: 'And the Vale of Siddim was full of slimepits; and the kings of Sodom and Gomorrah fled, and fell there . . .'.[1]

Natural seepages of oil, in pools, from shale and from tar sands, have been known since time immemorial and used from the earliest civilizations onward to provide bitumen for building, pitch for ships, glue for toys and unguents for arthritis, back pain and other ailments. Archaeologists have discovered that the Egyptians used oil to coat their mummy linens, making them especially vulnerable to fire hazard, and the craftsmen of Sumer to fix the ivory on their games boards. Herodotus records it as the source of eternal flame in Persia, a feature of the holy places of the Zoroastrians who worshipped the sun and fire. (Recent travellers to remoter northern Iran have encountered worshippers of Islam facing the sun rather than Mecca in their prayers.) The Byzantine general, Belisarius, drove a herd of animals covered in flaming bitumen into a terrified army of Vandals in AD 534, while his successors developed its use in the lost formula of 'Greek fire', the flaming projectiles hurled at the ships of Turks and Barbarians as they attempted to besiege the great city of Constantinople. Marco Polo mentions the oil trade in Baku on the Caspian Sea as early as the thirteenth century. Sir Walter Raleigh found the pitch lakes of Trinidad. Christopher Columbus came upon the North American Indians coating their bodies and mending their wounds with oil.

The use of oil by the Seneca Indians of Pennsylvania as a cure-all gave rise to the term Seneca Oil, or 'Snake Oil' as it became more familiarly known to locals and later *aficianados* of westerns. Samuel Kier, a Pittsburgh druggist, got hold of a source in Allegheny County and marketed it throughout the 1850s, selling it as 'Petroleum, or Rock Oil. A Natural Remedy, procured from a well 400 feet below the Earth's Surface'. Whether it worked or not, Kier managed to sell nearly a quarter of a million ½-pint bottles of the liquid without apparently incurring too many malpractice suits.

The original catalyst of the oil industry, however, was neither the rheumatism of the eastern United States nor the need for fiery missiles, but the rising demand for lighting in the nineteenth-century industrializing nations. In the United States and in Europe fish and whale oil, the traditional source of candle wax, could no longer satisfy the needs of industry. By the middle of the nineteenth century, the American whaling industry which had employed 738 vessels at its peak in 1848, was beginning to run out of raw material as overfishing decimated the whale schools and the ships of New England had to sail further and further out to find new sources. It also prompted a number of experiments on both sides of the Atlantic to see whether anything could be made of the shale oil and petroleum sources.

James Young

If any single figure deserves to be called the founder of the oil industry it is the Scottish chemist, James Young. One of those extraordinary, virtually self-taught and self-made men of Victorian Britain, he was the son of a Glasgow joiner. Working on the lathe by day, Young then went on to night school as well as attending chemistry lectures at Anderson's College, where he studied under the eminent Thomas Graham and made friends with the scientist, Lord Playfair and the African explorer David Livingstone. Livingstone he was to help support through most of his treks in search of the sources of the rivers of Africa. Lord Playfair introduced him to a natural petroleum source in the North Riding of Yorkshire, owned by his brother-in-law, when Young was working in Manchester. Young was quick to see the possibilities of refining the fuel as a lighting oil and when the Yorkshire source failed he moved to Glasgow where he patented a process and founded a company in 1851 to produce paraffin for lamps and lubricants from the local coal and shale oil. He had to spend much of the succeeding decade defending that patent against legal challenge and infringement. None the less Young's Paraffin Light and Mineral Oil Company provided the model for a rapid spread of oil-refining companies through Europe and the United States; where a Canadian geologist, Abraham Gesner, developed a similar process and dubbed the product 'kerosene' from the Greek words for oil and wax.

Colonel Drake and Titusville

There were more than a dozen refining plants in the United States and several times that figure in Europe when Colonel Edwin Drake started drilling for oil at Titusville nine years after Young patented his process. Drake himself was a somewhat unlikely figure to revolutionize an industry. Sick with neuralgia, he had ended a somewhat chequered professional life as a railroad conductor (a rather more important position then than now) when he was approached by a group of New Haven businessmen to invest in a failing company called the Pennsylvania Rock Oil Company. It had been founded by a young New York lawyer, George H. Bissell, who had been interested in the oil business since he had been persuaded on a visit to his old college at Dartmouth, that crude oil might produce a better quality of kerosene than oil refined the Young way out of coal and shale. Bissell then bought the source of the oil he had been shown in Dartmouth, which was at Titusville, and sought a second opinion from Benjamin Silliman jun., Professor of Chemistry at Yale. Silliman, who wisely insisted on being paid before he would hand over the results, gave the Titusville crude a rave review. 'In conclusion, gentlemen,' he declared, 'it appears to me that there is much ground for encouragement in the belief that your Company have in their possession a raw material from which, by a simple and not expensive process they may manufacture very valuable products.'[2]

What they couldn't manufacture, like many a company before and since, was enough start-up cash, which is where the New Haven businessmen and the 38-year-old Edwin Drake came in. Drake was sent to look over the properties and to manage them. Later, legend had it that he was chosen because, as a conductor, he could get a free pass on the railroads. The more likely reason was that he was at a loose end and had put $200 of his own money into the venture.

In any case, with $1000 from the company to set up the business, he went off to Titusville in early 1859. At first he did what the Indians had done before him and dug for the source of the oil that was collecting in a pool above the ground. It was slow work and he soon hit rock. He also hit on the idea that a much better system of exploration would be to copy what the locals did when searching for brine water by drilling for it. Erecting a 3·6-metre (12-foot) derrick, or tower, with an engine and pipe, he literally pounded the pipe down into the ground and through the rock. 'Drake's folly' they called it in the neighbourhood but after a summer's work of painfully slow progress, he finally struck oil at 21 metres (69½ feet) on 27 August. This was no gusher. The oil bubbled up with some water to be pumped into a vat at a rate of nearly 20 barrels a day.[3]

The Great Oil Rush

The Titusville discovery was the dawn of the oil era. It made the industry an extractive one, based on drilling. It also gave it its romance. 'The excitement attendant on the discovery of this vast source of oil', trumpeted the New York *Tribune*, one of the few newspapers to take note of the event, 'was fully equal to what I ever saw in California, when a large lump of gold was turned out.'[4] It started out as exactly that, a new gold rush attracting the same prospectors who had gone to California a decade before, bringing in the lawyers and the land title disputes, the whore houses and the gambling dens that were the hallmark of the gold-rush towns.

'God,' declared one local preacher, 'put that oil in the bowels of the earth to heat the fires of Hell. Would you thwart the Almighty and let sinners go unpunished?' The newly discovered oil, argued one of the ex-whalers, who had abandoned his old calling to join in the new rush, came from the blubber of a school of giant whales beached at the time of Noah's flood and subsequently buried beneath the earth. James Young, on a visit to wage a patents' battle, visited Titusville. Like many a businessman before and since surveying his competitor, he concluded: 'I dinna think it will come to much...' and that the new industry would never compete with Scottish shale oil.

He was wrong. With oil fetching $20 a barrel, the new source at Titusville, with a production cost of less than $1 per barrel, was extremely lucrative. The age of the oil bonanza had arrived. A forest of derricks sprouted across the landscape as speculators rushed to obtain leases and drill on any land they could get their hands on that seemed even vaguely connected with Drake's property. Land with the same above-ground features as Drake's, a bluff by the creek, fetched a premium, so did anything near it.

A new breed of 'oil smellers' came into existence, claiming to be able to detect sources of oil by smell. Others used water diviners and the visions given unto them in dreams. The new towns of Oil City, Petrolia, Babylon and, at the centre of it all, Pithole, sprang up. It was not only claims that were passed around the bars and pushed across the poker tables but also shares in those claims. Every waitress, reported a newspaper at the time, was spending part of her earnings on buying shares.

President Lincoln mentioned the new commodity as a source of vital strength for the North as it went to war with the southern states in 1861, as indeed it proved. The first exports started up from Philadelphia in December 1861, although the crew of the ship refused to handle the cargo when they found out that it was inflammable and the waterside dives had to be scoured for a rapid replacement. As the Civil War ended, Pennsylvania and the other states, where oil traces had been found, were filled with a fresh intake of army veterans wanting to make their fortune by either searching for it or speculating on it.

The American Civil War also saw the price and the fortunes of the infant industry waver madly as new fields were discovered and old ones exhausted. By the end of the war in 1865, California and Colorado had joined West Virginia, New York and Ohio as oil-producing states and drilling had started in Kansas, Tennessee and Wyoming. Many failed, including Drake's old Pennsylvania Rock Oil Company, its Titusville discovery worked dry. The post-war recession throughout the United States brought ruin to many of those who had made their fortunes out of the oil bubble. Colonel Drake, for a time Justice of the Peace for Titusville, lost heavily speculating on oil investments in New York. He died a virtual pauper in 1880, dependent on a $1500 state invalid pension handed out to him for his spinal neuralgia. Recognition came later in the form of a $100,000 monument erected in honour of the man who had founded the industry by a partner in the company which took most of the pickings, John D. Rockefeller's Standard Oil.

John D. Rockefeller

John Davison Rockefeller stands as such a giant of the industry, a man so hated and reviled in his day, that it is almost impossible to view him clearly, even now. His detractors saw him as a ruthless and cunning predator, a man who single-mindedly attempted to take over the entire industry in a passionless pursuit of power. His admirers – and they included not a few of his former competitors induced to become directors of the company which had destroyed them – saw him more as a man of daring and imagination with the vision and administrative ability to bring order to the new industry and to enable it to grow.

To those who met him, he seemed neither. A quiet, modest man with a shy smile and a directness of conversation, he sported neither the flamboyance of a Vanderbilt or J. P. Morgan nor the eccentricities and extreme views of a Henry Ford. A devout Baptist, he was thoughtful, correct and determined. His genius lay in the extent of his grasp of detail, his understanding of accounts and, above all, in his ability to reduce the problems of the oil business down to the essentials and to tackle and resolve them. The drama and contention were the result rather than the cause of his actions.

Rockefeller first came into contact with oil when he went out to Pennsylvania on behalf of his Cleveland firm to investigate the new industry at the height of the boom in 1860. He was 21 at the time, sharp, introverted and intensely ambitious.

Born in Moravia in upstate New York, his parents were straight out of a play by Eugene O'Neill or Strindberg. His father, German by descent, was something of a conman, a heavy-set extrovert with an eye for the ladies. His absences from home were frequent and his relations with his eldest son, John D., strained. At various times he is recorded as setting up as a quack doctor, advertising himself as 'The Celebrated Cancer Specialist' and as having been indicted for the rape of the family servant (although the charge was later dropped). The liberal Joseph Pulitzer, editor of the *New York World* and founder of the journalism prize that still bears his name, had his staff hunting for months for evidence that the old man was also a bigamist when the public outcry against his son was at its height at the beginning of the 1900s. He never did find any proof. John D. Rockefeller later recalled his father as having taught the lads to cheat at cards in order to bring them up 'sharp'. Although his father was to help finance his first business venture in Cleveland, it was strictly on the basis of receiving 10 per cent interest on the loan.

His mother, in contrast, was a determined, small-boned Scot of profound moral and religious persuasion; a member of the Baptist church who inveighed against violence and drummed into her six children the absolute virtues of thrift and hard work. She ruled the brood with a firm hand and, on her son's recollection, showed little open affection. Once, he wrote later, during a beating he managed to persuade her that on this occasion he was innocent. She proceeded with the task in hand with the comment, 'Never mind, we have started in on this whipping and it will do for the next time.'[5]

John clearly took after his mother both in looks and character, although it is too easy to be glib about these influences. Nor did the tensions and the lovelessness of his early years help. Throughout his life he was to shy away from public warmth, presenting to the world, in the words of his biographer Allan Nevins, 'a front of silence like smooth steel'.[6] 'Don't be a good fellow,' he reportedly told the pupils at the Sunday School at which he taught. 'I love my fellow man and take a great interest in him. But don't be convivial . . . don't let good-fellowship get the least hold on you. . . . It is my firm conviction that every downfall is traceable directly or indirectly to the victim's good-fellowship, his good cheer among his friends, who come as quickly as they go.'[7]

He was, and remained until his death, a prominent and committed member of the Baptist church, with an instinctive dislike of display and a precise view of charity. Early on he kept an account book detailing his donations. They show him giving a sizeable proportion of his first salary ($9.09 out of the $95 he had earned for his first four months) to a wide range of causes from the Sunday School to runaway slaves and the poor. He was to keep the habit all his life, while continuing to express the parsimony sometimes shown by the very wealthy.

'Attention to little details', was the formula for success that he gave to the many who asked him. It was typical of him to regard it both as his pious duty to give and to account for it in ledgers. He loved figures and studied book-keeping at night school while earning money as a clerk during the day. At 18 he set up a commission agency on the Cleveland waterfront with a friend, an immigrant from England called Maurice Clark. It was good timing. Commission agents took a straight commission on buying or selling goods and the Civil War and the years immediately preceding it enormously boosted the demand for a whole range of items in the North. By the end of the first year Rockefeller and Clark had collected commissions on a gross turnover of $½ million.

Rockefeller's trip to the Pennsylvania oilfields in 1860 was presumably in pursuit of new commodities to factor. As has been mentioned, he was not impressed at what he saw. Production and exploration in the oilfields was completely uncontrolled and prices, already halved, reached rock bottom the following year. The sight of wells running dry filled Rockefeller with dread at the dependence on so uncertain a source of supply and one so subject to price fluctuation.

It was, instead, through refining that Rockefeller came into the industry three years later and it was the manufacturing part of the industry that served as the fulcrum of his business. Cleveland had by that time become something of a refining centre for the American northeast. It was conveniently situated along three major rail networks from Pennsylvania on the banks of Lake Erie and the Great Lakes' complex that led to the eastern seaboard. The cost of refining the new crude oil was but a fraction of that of refining shale oil to make kerosene. A crude-oil refinery, basically little more than a pressure vessel to heat the oil and separate out the heavier and lighter oils from the desired kerosene, cost anything between $200 and $4000, depending on size. The equivalent coal- or shale-refining plant would cost up to $100,000. Within six months of Drake's discovery, fifteen new crude-oil refineries had been built in the region, several of them in Cleveland. By the mid-1860s the number had doubled in Cleveland alone, and nearly quadrupled in Pittsburgh.

In 1863 Rockefeller and Clark took a share in the refining business of Samuel Andrews, a fellow immigrant from the same town in Wiltshire as Clark. Andrews, a former candlemaker who understood lighting, had the reputation of manufacturing more kerosene (about 70 per cent) from a barrel of Pennyslvania crude oil than anyone else and producing a product that, because it contained less gasoline, was more stable when it burned.

When Andrews decided to expand in 1865, at the end of the Civil War and Clark, who was doing very nicely out of the commission business, proved reluctant to put in more money, Rockefeller decided to put all his eggs into the oil basket. He exchanged his shares in Clark and Rockefeller for Clark's shares in the Andrews' refinery for which he also had to pay out $72,500, an indication of how big a step into the unknown he was taking. Always good with bankers, Rockefeller got the money and started the new company, this time with his own name in front, Rockefeller and Andrews. When he needed still more money to expand two years later, he brought in Henry M. Flagler, a fellow upstate New Yorker who had left school at 14, worked along the Erie Canal and had then had the good fortune, or skill, to marry the niece of the Cleveland whiskey-distilling magnate, Stephen V. Harkness. Harkness put in two-thirds of the $90,000 needed at the time. Flagler, a man of immense energy and considerable vision, was to be Rockefeller's closest associate in the first decade of growth and was to go on in the 1880s to promote Florida, financing and organizing the building of railways and hotels, the dredging of Miami Harbour and the establishment of shipping lines.

Rockefeller set about developing his refining business in a characteristic manner. A tight rein was kept on costs and he moved, in the classic late nineteenth-century manner, to control production whereever possible, setting up his own cooperage to make the barrels in which the products were shipped out and developing his own local distribution. Instead of using jobbers, or wholesalers, to buy his crude oil as was normal practice for refiners, he bought direct from the oil producers, although the production end was too chancy a business to tempt him in himself.

In 1870, to ease the problems of funding its expansion, the company was turned into a joint-stock company and renamed the Standard Oil Company. Incorporated in Cleveland, it had John D. Rockefeller as its president, his brother William, who acted as his export manager and east-coast representative in New York, as vice-president and Flagler as secretary and treasurer. Andrews did not appear on the masthead, although he remained head of refining. He left the firm in 1878, a technician out of place in the empire Standard was by then becoming. The name Standard was chosen to emphasize the consistency of the product. It might also have emphasized the consistency of Rockefeller's determination to expand. With refining, like production, the victim of too many entrants, too much cost cutting and too little organization, Rockefeller and Flagler set about a strategy of development based on low costs, high through-put and the destruction of the competition through take-over, price cutting and aggressive marketing.

It was the kind of business that suited Rockefeller's talents. Attention to detail was what mattered above all. When, so a later story went, Rockefeller was watching kerosone cans being soldered shut, he asked the supervisor how many drops of solder were used. The figure forty was given and Rockefeller suggested they try thirty-eight. When that proved not quite enough, the company was asked to find out the correct number. Thirty-nine was the answer and thirty-nine became the rule.[8]

'No Oil for the Ring'

Refining costs were certainly one area for control. The most critical element for a Cleveland refinery was, however, transportation, both of crude oil into the refinery and kerosene out of it to the cities of the east and the export ports. To grow, Rockefeller needed to expand his intake and his output and for this he needed the cheapest possible rates from the railways.

A hundred years on it is difficult to envisage the power of the railways and the rail magnates of the time. It was they who controlled the pace and sometimes the direction of the expanding industrial United States. A town on the railway became part of that expansion; by-passed, it would be relegated to a backwater. They were, in the manner of nineteenth-century barons, both highly competitive and quick to form cartels to sustain rates when the competition proved too damaging.

Transportation from the oilfields and from Cleveland was dominated by three major systems: the Erie, run by Jim Fisk and Jay Gould; the New York Central, the creation of Commodore Vanderbilt, famous for his constant refrain 'the public be damn'd'; and the Pennsylvania, run by Thomas Scott. In 1868 Rockefeller was able to use his volume to obtain rebates of 10 to 15 cents per barrel on shipments in and out of Cleveland, enough to give him the added edge to force out many of his competitors.

In 1872, with a widespread recession gathering pace (it was to prove as bad in its way as the 1930s' Depression), Rockefeller went much further. The railway magnates, seeing many of their customers in industry as well as oil, going bankrupt or merging, came to Rockefeller and Flagler with a proposition. The three railways and thirteen major refineries in Cleveland should each form themselves into associations. The railroads would share out the business between themselves. The refiners would get a discount not just on the oil they sent but on any oil others sent. The independent refiners would just have to lump it.

Doubtful at first, Rockefeller was determined once he had agreed and the South Improvement Society was formed by the refiners. A rate of $1 per barrel was agreed instead of the normal 80 cents per barrel of crude from the fields plus $2 per barrel of kerosene from Cleveland to New York. New and higher rates would be introduced for everyone else. The members were sworn to secrecy, both about the prices and the negotiations. When the news leaked prematurely it caused consternation in the oilfields and fury in Cleveland. The oil producers were particularly incensed. Such a combine of refiners threatened not just their outlets among the independent refiners but could have led the way to the formation of a cartel of crude-oil buyers as well who could force down the price. Titusville opera house was taken over for a meeting of more than 3000 protesters shouting, 'Down with the conspirators.' Another protest meeting was held in Oil City. An association of producers was formed to meet the new challenge. Called the Petroleum Producers' Union, its secretary was a young independent refiner, John Archbold. Rallying round the slogan 'No Oil for the Ring', they swore not to sell a barrel of crude to the members of the South Improvement Society and stuck to their resolution for over a month.

The producers won. The press was on their side. Public opinion was outraged. The Pennsylvania legislature debated rescinding the South Improvement Society's charter altogether following a committee investigation which concluded that it 'was one of the most gigantic and dangerous conspiracies ever conceived'. In a scene worthy of their surroundings, representatives of the Producers' Union met the leaders of the railroads at the Grand Opera House in New York. Rockefeller and his colleagues from the South Improvement Society were barred entry. One by one the railroads capitulated.

The incident marked Rockfeller for ever in the eyes of the public with the stain of the greedy monopolist and signalled the beginning of the struggle between oil producer and oil refiner/marketer which has dominated the industry since. In fact he was never the leader of the conspiracy, although he was one of the toughest stand-outs once the confrontation had begun. Nor did he need such a formal cartel to achieve his aims. Volume of production and good management allowed him to undercut most of his competitors on even ground. Anyway, the recession had helped to wipe out most of the opposition. By the following year the number of refiners in Cleveland had been reduced from thirty to ten and Rockefeller was busily expanding outside the region. His major battles now were less against competitive refiners – where his edge had been increased by developments in technology which required much higher investment, which he could afford to pay for, than the early crude distillation vessels – than with the railroads and the producers over distribution.

After the defeat of the oil war of 1872, the Pennsylvania Railroad built up in earnest its oil transportation subsidiary, the Empire Transportation Company, as part of its plan to dominate the movement of oil to the east coast through a complex of pipelines, rail-tank cars, steamers and tank storage facilities. Rockefeller retaliated with a ferocious price war on any refiners who used Empire's facilities. By 1877 Pennsylvania was reported to be losing around $1 million a month and it had to pass on paying its dividend for the first time ever. It was also, coincidentally, hit by violent strikes. Exhausted by reverses on all fronts, Thomas Scott sued for peace and, like so many of the refiners who battled against Rockefeller, sold to his rival the whole Empire oil transportation subsidiary at barely more than book value.

Rockefeller now ran about 90 per cent of the Cleveland refining industry. When he added the last remaining independent pipeline to his stable, the National Conduit Company, he also gained control of the distribution of crude oil. The producers had to react. They besieged the Pennsylvania Government to take action that would force Rockefeller to move their oil at fair prices. They pursued both the railroads and Rockefeller in court and helped to instigate a series of inquiries into his methods and his use of rate rebates to force out competition. They even planned to build a long-distance pipeline of their own before a truce was finally called and Rockefeller and his allies in the railroads promised fairer terms.

In 1879 a group of entrepreneurs, using the engineering skills of Civil War veteran, General Herman Haupt, succeeded in an even more daring and potentially challenging scheme with a plan to build a 100-mile pipeline across the Allegheny mountains to the coast at Williamsburg, using pumping stations to push the oil along the way. The company, the Tidewater Pipe Company, was backed by the independent producers. At first few took it seriously. Pipelines were used for short-distance transportation, mainly to the railheads. It simply did not seem feasible to pipe over such a distance, let alone over a mountain range. But it was, and the line opened later that year, refusing from the start to accept any oil from Rockefeller. Standard Oil tried every trick in the book to defeat the new system, including setting up shareholders to petition the courts to have Tidewater put into receivership. The case was dismissed. Undaunted, Rockefeller built a long-distance line of his own, started a rate war and finally wore Tidewater down to the point where it was forced to accept a loan from Standard. The independent producers who had founded Tidewater – Byron D. Benson, Robert Hopkins and David McKelvy – were voted off the board and a new board, sympathetic to Standard, was voted in. An arrangement was reached by which Standard's subsidiary carried 88·5 per cent of all the oil out of the Pennsylvania fields and Tidewater took the remainder.

The Standard Oil Trust

Tidewater's defeat effectively sealed Standard Oil's dominance of the region. Rockefeller's energies were becoming increasingly directed to the national and international stage. His headquarters were moved to New York and he himself went to live there at what is now the Rockefeller Center. The major problem he now faced was a legal one. Under the law of most American states a company registered in a particular state could not own companies elsewhere. This had not stopped Rockefeller from buying out or setting up oil-refining and transportation companies all over the United States. The shares in these companies, however, had to be held by individual directors as trustees and it was impossible to consolidate the structure into an orderly framework. Until, that is, Standard's chief lawyer, C. T. Dodd, formed the Standard Oil Trust in 1882. By means of this legal device, quickly copied by a large number of industrial conglomerates eager to expand in the same way, all of Standard's associates and subsidiaries were reorganized into separate companies in each state, their shares vested in a central board of trustees, presided over by Rockefeller, in return for trust certificates entitling them to dividends but not voting powers.

Strictly speaking the trust lasted for barely ten years for in 1892 the Supreme Court of Ohio ordered its dissolution on the grounds that it offended against the company's original charter. Appealing against the ruling, Rockefeller delayed matters until he was able to take advantage of a change in New Jersey law that allowed companies registered there to hold shares in companies in other states. Standard Oil (New Jersey) was born as the holding company for the Standard empire.

To Rockefeller this was just good business sense. An industry like oil could not be developed to its maximum potential and its constituent parts brought together in a properly organized form unless it was through centralized control and standardized product. The establishment of the trust, he later recalled 'was the origin of the whole system of modern economic administration. It has revolutionised the way of doing business all over the world. The time was ripe for it. . . . The day of combination is here to stay. Individualism has gone, never to return.'[9]

This was only partially true. Although syndicates were to become the style of big business in the United States through the turn of the nineteenth century into the twentieth, individualism still ruled the popular imagination. It had not entirely disappeared and it was to return. As Standard Oil spread its hold on the oil business, from refining and transportation to distribution and sales, and as the rise of the automobile industry broadened the scope of the business considerably to provide oil for transport as well as for heating, so the outcry against Standard's dominance grew.

Rockefeller under Attack

First Henry Demarest Lloyd, a leading writer in the anti-big business *Chicago Tribune* and the *Atlantic Monthly* and then, with even more effect, the writer and academic Ida M. Tarbell, launched into popular print with a series of scathing exposés of Rockefeller's business methods, accusing him of driving out his rivals by force of unfair competitive practices as well as outright bribery and corruption. Lloyd, known as the 'father of the muckrakers', assaulted Standard Oil in articles and then in a bestselling book, *Wealth and Commonwealth*, published in 1894, accusing the company of being run by the new barbarians who would destroy civilization. 'To them science is but a never-ending repertoire of investments,' he wrote, 'stored up by nature for the syndicates, government but a fountain of franchises, nations but customers in squads.' His attacks at the time seemed to be based on the fact that the new industrial moguls like Rockefeller lacked breeding and culture as much as they lacked moral scruples.

Ida Tarbell, hired by the muckraking *McClure's* magazine to investigate Rockefeller's empire at the turn of the century, developed her attack from a more romantic standpoint. Her father had been an independent oil producer and a leading member of the Producers' Union. She remembered the days of the oil rush as a halcyon era of rogues, reprobates, entrepreneurs and chance-takers, men who were, above all, individualists. And she brought cold anger to her descriptions of the methods by which Rockefeller had crushed them.

Some of these attacks, pursued with vigour in William Randolph Hearst's New York *Journal* and his other papers as well as in Pulitzer's *New York World*, were somewhat unfair to Rockefeller, the whines, as he saw it, of people who had been forced to sell or go out of business through their own inefficiency rather than his machinations, producers whose wells had run dry for geological reasons, or whose plants had become uncompetitive during the frequent periods of recession or falling prices. To the end of his days, Rockefeller, encouraged by his aides, believed that his fault lay not so much in what he had set out to do but in his failure to understand and exploit his public relations. If only he had brought the journalists into his confidence at the start, he rationalized later, the misunderstandings would not have occurred.

This was simple blindness to the reality of the situation. Whatever the truth of the individual stories that Tarbell and Lloyd raked over – of widows pressurized into selling their shares, or refiners battered into submission and forced to sell their plant at less than the cost of building it, of independent oil producers starved into bankruptcy for their refusal to pay the exorbitant fees demanded by Rockefeller's pipeline companies – the fact of the matter was that the oil industry was growing too big and too fast to be controlled by a single enterprise. Rockefeller's rationalization may have been essential, or useful at any rate, in moving the industry on from its first phase of boom-and-bust competition as thousands flooded into the oilfields, hundreds of small refineries were erected and the railways tried to maintain a stranglehold on transportation, but it couldn't last once his own dream of seeing oil become a national industry came true.

Rockefeller's instinct – and the instinct of Archbold and others around him – was to keep control always, well beyond the level of what could be justified to achieve economies of scale or rationalization of facilities. The refineries and companies outside the Cleveland region taken over by Standard during the 1870s, 1880s and 1890s – Continental Oil Company, Illinois; Vacuum Oil Company, New York; Sun Oil; Carter Oil Company; and Kendall Refining Company to name some of the more prominent – gives a fearsome indication of how much of Standard's growth came from acquisition rather than self-generation.

By the turn of the century Standard Oil was also developing much further both into production and direct retailing. In the 1880s Rockefeller's continual nightmare, that the wells of Pennsylvania would start drying up, actually started to occur. In the mid-eighties production switched more and more to the new discoveries in Ohio and Indiana where the huge Lima field had been discovered. The oil from Lima was, however, high in sulphur. Against the advice of Archbold and the other directors, Rockefeller decided to go directly into production and to pour money into research to find ways of overcoming the problem. He was successful in both. By the time public outcry finally induced Congress and the local state legislatures to begin hard investigations into Standard, its relationships with the railroads and its commercial practices, Rockefeller admitted that his company controlled something like 90 per cent of the entire industry in the United States.

Spindletop

Such a state of affairs couldn't last. Exploration, as so often in the history of the industry, broke the stranglehold in the form of the great gusher at Spindletop in Texas. It was the achievement of one of the oil industry's genuine visionaries, Patillo Higgins, a one-armed ex-lumberjack turned property speculator who became convinced that a giant oilfield lay beneath the sulphurous mound of Spindletop outside his home town of Beaumont. Returning home from a wandering life, Higgins – incidentally, like Rockefeller, a teacher in the local Baptist Sunday School – began the search for oil and dreamed of creating a model city from the proceeds on top of the site. For eight years he drilled away on the mound, where gas seepages had convinced him that oil lay beneath. The geologists disagreed. His local backers grew dispirited and his drilling never quite managed to go deep enough. In 1899 he was forced to advertise for a partner and sold his operation, in exchange for an interest, to Captain Anthony Lucas, an engineer in the Austrian Navy who had spent some years salt-mining in Louisiana and was sure that the sources of salt and oil were connected. Living on a shack atop the mound, he too exhausted his savings over months of drilling.

Finally Lucas sought the help of Standard Oil. Calvin Payne, the company's production expert, was sent down and, after surveying the landscape, which was unlike any other in which oil had been found, he reported back that there 'was no indication whatever to warrant the expectation of an oilfield on the prairies of southeastern Texas'.[10] Undeterred, Lucas turned to Dr William Phillips, Professor of Geology at the University of Texas, who was more encouraging. Armed with his report, Lucas brought in two new partners, the Pittsburgh oil prospectors Guffey and Galey. The money asked for was large by the standards of the day – $300,000 – but the partners eventually agreed to come in with Lucas on the sound condition that they did the drilling and that one deep hole only be tried. If it worked, all well and good. If not, the operation would be closed down. The added requirement was that Lucas buy as much land as possible in the vicinity.

The drill used was a rotary one, a recently developed technique which involved screwing through the earth and rock, quite unlike the Pennsylvania system which virtually pile-drove through it. It succeeded. After six weeks of drilling for eighteen hours a day, coping with a gas blow-out and seeing half their men leave in despair, the partners finally struck oil on 9 December 1900. They drilled further, pushing to a depth of 305 metres (1000 feet) on the first day of 1901. On 10 January, with a new bit freshly installed, the whole well erupted. With a whooshing of gas, drill bits and equipment were thrown aside as a great column of oil exploded 60·9 metres (200 feet) into the air. 'Thank God,' said an unbelieving Lucas.

Spindletop was the true Hollywood version of the gusher. The well flowed 80,000 barrels a day, half the total oil production of the United States at the time. Sadly it was not to have a Hollywood ending. Lacking the capital to control a discovery of this magnitude, Colonel Guffey was even forced to seek an emergency loan for the equipment to hold the oil gushing out of the well and then to get a capital injection of $3 million to develop the field. He got them – from the Mellon bank in Pittsburgh – but at a price. Both Galey and Lucas were to be bought out and the company reorganized with Mellon in control. As in Pennsylvania, the pioneers gained little glory, let alone financial reward, for their efforts. Higgins had to go to court to get even a settlement for his share and died in poverty aged 92 after half a century of fruitless search for other bonanzas. Lucas went off to Washington and Colonel Guffey, now tied to the Mellons, was forced out of his own company in 1907.

Worse was to befall Higgins' dream of a model city. Within days of the discovery, an oil rush started. Surrounding land was snatched up at grotesque prices. Wells were drilled on every likely spot, including Spindletop itself where over 400 had been made by the end of the year, creating a forest of derricks within feet of each other. The pressure went from the reservoir and the wells stopped gushing. By the time a year was out production was down to 5000 barrels per day, a sixteenth of the output of the first well. Captain Lucas came down to survey the devastation of abandoned derricks, rusting equipment and polluted ground and commented sadly: 'The cow was milked too hard and moreover she was not milked intelligently.'[11] It was one more image that stayed with the industry through its development.

Spindletop was, however, not just a flash in the pan. Its success encouraged a wave of drilling in the American southwest, where the prairies – contrary to Standard Oil's initial judgment – produced a ceaseless stream of oil discoveries and created new sources of oil outside Standard's reach. Indeed, the Texans deliberately cut Standard Oil out wherever possible, inducing a company spokesman to declare that, '. . . after the way Mr Rockefeller was treated by the state of Texas, he'll never put another dime down there'.[12] The Texan and neighbouring discoveries also created powerful new companies which were prepared to take on Standard Oil and survive: Mellon's Gulf Oil, developed from the company formed in Guffey's name in 1907; the Texas Company, later Texaco, founded by J. S. Cullinan; Humble Oil; and Houston Oil.

New oil sources met increasing demand as the automobile, at first slow to take off, its growth restricted by its cost and technical limitations, became the product of the new century. In 1900, 4000 automobiles were sold in the United States, doubling in one year the total number in the country. By 1910 the number of automobiles produced there was up nearly fifty times at 187,000. At the same time, oil was increasingly substituted for coal to drive marine engines and the railway locomotive, as well as to power the machinery in factories. The whole field of transport opened up within a generation at the beginning of the twentieth century. Oil, gradually being displaced in the lighting market, was becoming instead the fuel of the new age of mobility. Gasoline was distributed in the same way as lighting kerosene, but on a larger scale. It was sold in cans through large numbers of local agents, with distinctive brands and a strong sense of user identity. Advertising and free gifts became part of the marketing battle. The world of Esso's 'Put a Tiger in Your Tank' and Shell's 'Go Well with Shell' was beginning to appear on the horizon.

So was the complex world of oil refining and marketing as we know it today. Crude oil, when heated in a simple refining process, naturally breaks into a range of liquids of varying weight. At the lighter end there are gasoline, naphtha (used to manufacture chemicals) and petroleum gases. In the 'middle of the barrel' are the 'distillates': kerosene, gas and diesel oils. At the heavier end are the fuel oils used in bulk heating, while at the very bottom is the bitumen used in roads and construction, lubricants and detergents. So long as the market for oil was limited largely to sales of kerosene for lamps, much of the crude oil was wasted and the naturally heavier oils discovered in California and parts of Asia were of little value. Once markets were developed for the full range of oil products, the industry became much more flexible and whole new vistas of growth and use opened up for both the producer and the refiner. The United States led the way. Europe, where coal was still so cheap, was slower. Even so, in Britain, the largest European market, the market expanded from 5 million gallons in 1900 (over 80 per cent of it kerosene) to nearer 500 million gallons in 1913, of which 20 per cent was gasoline and 20 per cent fuel oil. Sales of fuel oil had dropped from 215 million gallons to 157 million in the meantime.

Anti-Trust and the Break-up of Standard Oil

Standard Oil attempted to keep up with this pace in the United States, buying into these discoveries and switching its refining and marketing strategy away from kerosene to gasoline, as automobile fuel has always been called in the United States. It was having difficulty sustaining its hold on the industry – the pace of expansion in demand was too much even for its resources, while the geographical variety of new discoveries was dramatically altering the pattern of crude-oil supply across the country – when the American legislature finally struck.

The widespread adoption of the Standard Oil style of trust in the 1880s by companies in the whiskey, sugar, rail and manufacturing industries had quickly brought down on business a response in the form of the Sherman Anti-Trust Act passed in 1890, the basis of all subsequent anti-trust legislation in the United States. Standard Oil, as the progenitor of this hated form of power concentration, and Rockefeller, now the symbol of the secretive and devious pursuit of monopoly, was the subject of consistent investigation in both Ohio and Washington. For a time he was able to avoid such attacks through his New Jersey holding company. By 1906, however, action was being taken against Standard companies in more than twenty states and the Attorney-General, under criticism for his failure to put bite into the Sherman Act, announced that he would bring suit against Standard Oil and its thirty-three subsidiaries. The case dragged on but in 1909 the courts decided against Standard and ordered that Standard Oil (New Jersey) cease ownership of the various companies. Finally, in a 20,000-word ruling of the Supreme Court in May 1911, Standard Oil was held to be in breach of the anti-trust legislation and all its subsidiaries and associated companies were ordered to be reconstituted as fully independent operating companies within six months. The great Standard empire was broken up, although the shareholders were given stock in all the companies pro rata, thus continuing the ownership pattern for a time.

The pressures of expansion and competition, added to that of the new legislation, increasingly pushed the children of Standard Oil in

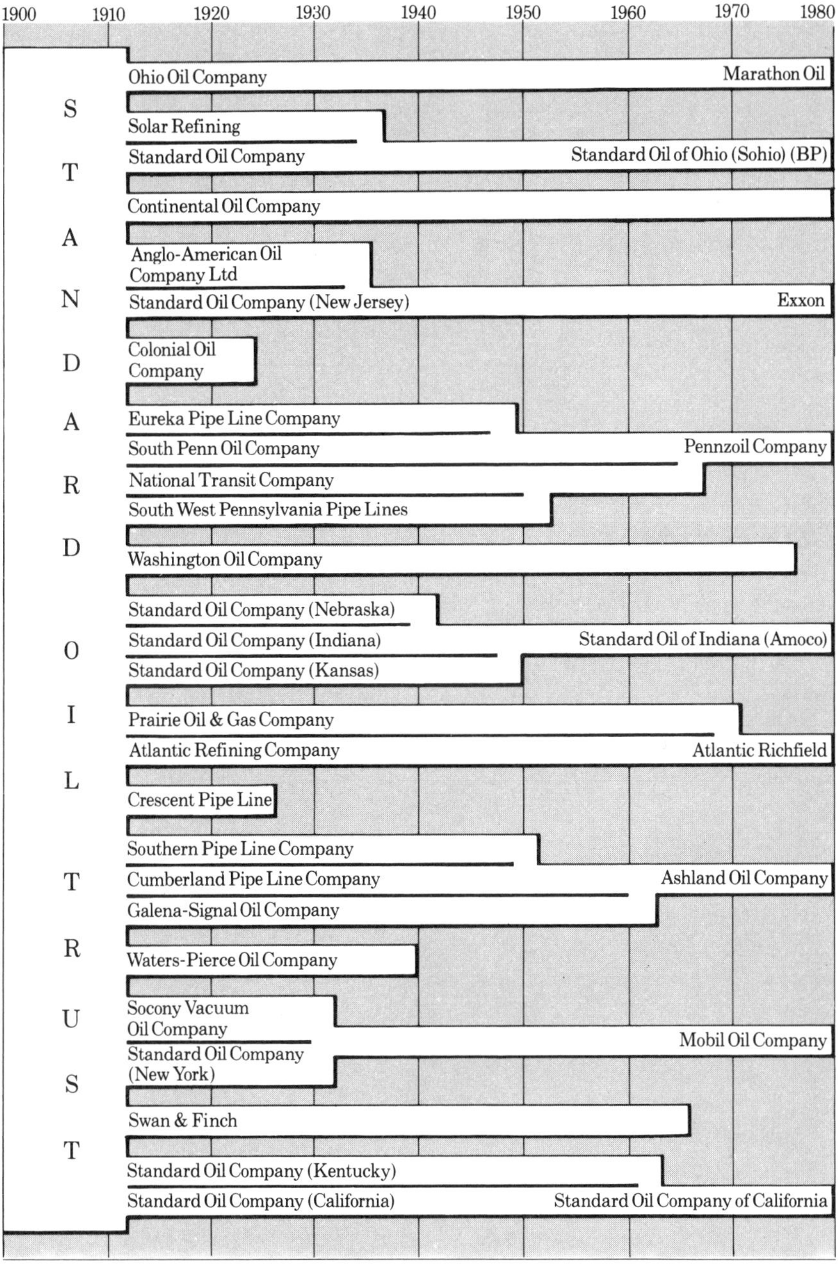

1900 1910 1920 1930 1940 1950 1960 1970 1980

STANDARD OIL TRUST

Ohio Oil Company
Marathon Oil
Solar Refining
Standard Oil Company
Standard Oil of Ohio (Sohio) (BP)
Continental Oil Company
Anglo-American Oil Company Ltd
Standard Oil Company (New Jersey)
Exxon
Colonial Oil Company
Eureka Pipe Line Company
South Penn Oil Company
Pennzoil Company
National Transit Company
South West Pennsylvania Pipe Lines
Washington Oil Company
Standard Oil Company (Nebraska)
Standard Oil Company (Indiana)
Standard Oil of Indiana (Amoco)
Standard Oil Company (Kansas)
Prairie Oil & Gas Company
Atlantic Refining Company
Atlantic Richfield
Crescent Pipe Line
Southern Pipe Line Company
Cumberland Pipe Line Company
Ashland Oil Company
Galena-Signal Oil Company
Waters-Pierce Oil Company
Socony Vacuum Oil Company
Standard Oil Company (New York)
Mobil Oil Company
Swan & Finch
Standard Oil Company (Kentucky)
Standard Oil Company (California)
Standard Oil Company of California

Source: *Fundamentals of the Petroleum Industry* by Robert O. Anderson. Copyright © 1984 by Robert O. Anderson.

The descendants of the Standard Oil Trust. Broken up into thirty-four companies by a historic judgment of the Supreme Court in 1911, most of the subsidiaries thrived, rivalling the size of the original trust. Three became 'majors' – Exxon, Mobil, and Socal – and seven became leading independents, to rank in the top fifty industrial companies of the United States.

different directions. The names remained largely the same but the regional pulls were different and by the 1920s most of them were branching out into each other's territory in direct competition. Of the three dozen companies which made up the original grouping, a dozen thrived, each becoming in time as big in financial terms as the original whole. The biggest, Standard Oil of New Jersey, or Exxon as it is now called (Esso in Europe), remains the largest oil company in the world, dwarfing most other companies in the United States. The Standard Oil Company of California (Socal), itself a merger of the Standard Oil companies of California and Kentucky, and Mobil, a conjunction of the Standard Oil Company (New York) and another Standard company, the Socony Vacuum Oil Company, both took their position among the seven major international oil companies, the 'Seven Sisters' of the post Second World War period. Among the top twenty oil companies of the United States along with the Seven Sisters today are: Marathon Oil, derived from the Ohio Oil Company; Sohio, from the Standard Oil Company of Ohio and now partly owned by British Petroleum (BP); the Continental Oil Company (Conoco), now taken over by DuPont; Standard Oil of Indiana, or Amoco as it is now called, from the Standard Oil Companies of Nebraska and Indiana; Atlantic Richfield, from Atlantic Refining and Prairie Oil and Gas; and the Ashland Oil Company, from Galena-Signal Oil, Southern Pipe Line and Cumberland Pipe Line.

The Rockefeller Achievement

John D. Rockefeller himself had been retired fourteen years by the time of the Supreme Court decision, although he was brought back several times to give evidence to the courts and congressional committees. He had been replaced, with great success, by John D. Archbold. Rockefeller had become too exposed and too infamous a name for the corporation as it became increasingly embroiled in defending itself against political attack. In addition, Rockefeller's style of management – the attention to detail, the tightness of control and the pursuit of cost cutting – was becoming less relevant to the needs of the developing industry.

His methods had worked supremely well when the core of the industry was refining and control of supply. They worked less well as expansion became too rapid to be controlled in this way. The automobile era brought in the arts of the marketer and the advertiser who could appeal both to the individual customer and the masses. The spread and shifting fortunes of exploration required more the skills of the dealer, the negotiator and the chance-taker. Rockefeller was none of these things. He was an organizer and a consolidator, who indeed gave order to an industry that badly needed it to find the right basis for international and national growth. However, the methods he used to run the industry were to turn into some of its most hated characteristics – its sheer size, its integration through the various phases of the business, its tendency to undercut opposition as a means of removing competition, its taste for the behind-the-scenes deal and, above all perhaps, its lack of accountability to individual, national or local political authority. The Rockefeller era personified these themes: i.e. the conflict between the individualism of oil explorer and independent refiner and the bureaucratic organized corporate structure required to move oil through the system to market; the triumph of lawyer, banker and dealer over visionary and pioneer; the gulf between the self-reliance and secrecy of the firm and the openness demanded by press and politicians.

Rockefeller genuinely could not understand why he was considered an ogre and his Standard Oil Company an 'octopus'. Most of the outside world could not understand how he could have possibly countenanced and pursued the elimination of all competition so effectively as he did without being an ogre. There was far more of the nineteenth-century industrialist, without the vulgarity, in this than the twentieth-century forerunner. In a sense Rockefeller pioneered the modern multinational corporation by founding a multi-state oil company in the United States. But in another sense he had far more in common with the manufacturers of the late nineteenth century than the corporate men of the twentieth.

Retiring to golf and his New York estates, he eventually outlived much of the calumny that preceded the Supreme Court dissolution of his empire. He lived to be 98 and by the time of his death in 1937 he was regarded more as a philanthropist and the survivor of a bygone age than a ruthless predator. Historians had been hired to put his achievement in proper perspective. Public relations experts had been taken on the pay roll to put gloss on his humanity, to give him some warmth as a family man, the old gentleman who strode out to golf until the last, with an extra pocket in his trousers filled with dimes to hand out to anyone and everybody, before the cameras as much as off them.

The persona was not entirely fictitious. Rockefeller was never what he hated most – 'a good fellow' – but he was a great philanthropist. He handed on his holdings in Standard Oil to his children and his trusts, although he kept share certificate number one as a keepsake. He multiplied his wealth and, over forty years of retirement, gave away $550 million of it. His only son gave away $400 million more by the time he died in 1960 at the age of 86. Nelson Rockefeller and his other grandchildren continued the tradition until now, when the wealth of the family and the Rockefeller Center in New York look as if they will finally be broken up and divided among the great-grandchildren.

'I don't feel the slightest bit apologetic for my grandfather,' argues David Rockefeller, who knew him well in his declining years. 'I think that history will indicate that what he did for the country in building the oil industry was a very major contribution to the success of our country as a major industrial power. I'm also very proud of the fact that, having made a very large fortune, he had the . . . I would say, genius . . . to give that money away for some of the most creative things that have been done in the field of philanthropy. He started in 1901 the Rockefeller Institute for Medical Research, which is now the Rockefeller University. He started the University of Chicago even before that, which has become one of the great universities of the world. Another quite remarkable thing is that, back before the turn of the century, he started a school for black women, now called Spellman College. It still exists in Atlanta, Georgia. It was long before the present concern for equality among the races. Furthermore, it was long before women were given an opportunity.

'I think the criticism of my grandfather,' David Rockefeller adds, 'was due in large part to the fact that his own success was in his ability to pull together a very disorganized and disunited industry which really was going nowhere until he organized the Standard Oil Company and pulled together the different elements in it.

'It's also true that, in the process of doing that, he used some quite ruthless techniques, some of which would be illegal today; but most of which at the time were taken for granted and were perfectly acceptable, at least legally. And many people did not think there was anything wrong about them. However, he did force a number of his competitors to sell out to Standard Oil Company. If one goes back and looks at the story and sees what happened to those competitors who sold out, those who accepted Standard Oil's stock, which he always offered, did very well. Those who accepted cash, as was the case of Ida Tarbell's father, they might not have invested it so wisely and probably didn't do so well.'

No one could doubt that Rockefeller himself had invested well.

2

Floating to Victory

The struggles of the early American oil industry were repeated with equal intensity on the international arena. Standard Oil was just as big and as threatening an influence abroad as at home. By the 1880s no less than 70 per cent of its production was going abroad, almost entirely in the form of kerosene and lubricating oils, to distribution bases throughout most of the world. It dominated the Asian market and gained a goodly share of the European market as well – 'carried wherever a wheel can roll or a camel's hoof be planted', as its advertising went – until challenged by a company as big and as ruthless as itself.

As the oil industry became international, the role of oil in national security became increasingly important. To the Americans, blessed with self-sufficiency in oil from the start, the business was about supplying kerosene and then gasoline for the ordinary customer with some lubricating oil and fuel oil for industry emerging alongside in the early twentieth century. The central political issue was one of monopoly, the corporation versus the individual. The Europeans were much less concerned with this. Protection and monopoly had always been part of mercantile history. Oil had another aspect to it. By the First World War it had become the vital artery of victory, the fuel for ship and tank. Controlling the sources of this fuel had, equally, become a part of the strategic thinking of Europeans locked in struggle for dominance of the continent and of their overseas empires.

It posed a considerable problem. For oil, unlike the coal that had fuelled the Industrial Revolution in Europe, was not found within the boundaries of the major Western European nations. Although individual finds were made in Southeast Asia, the Middle East and Eastern Europe and although Germany, France and Britain all invested heavily to develop the manufacture of kerosene from local coal and shale sources, Europe remained heavily dependent on the United States for its supplies until the 1870s when the extensive reserves of Baku in Russia started to come on to the market.

Baku and the Nobel Brothers

Baku proved to be to the eastern side of the Atlantic what Pennsylvania had been to the western side. Situated on what had originally been Persian territory, until ceded to Russia at the beginning of the century, Baku was on the Caspian, on whose surface seepages regularly accumulated to be set alight for the amusement of important visitors. Development was at first held back by the fact that the mineral rights of the area, held formally by the Tsar, had been ceded to a merchant called Meerzoeff as a monopoly in 1861. When the Russian Government ended this arrangement and started to auction plots of land, activity exploded. So did the discoveries. Within a few years, despite its distance from the main rail or sea transport systems, Baku had become a full-throated frontier boom town. The wells were far more prolific than those in Pennsylvania. They were called spouters since they threw sand and earth several hundred feet up and, unless quickly controlled, spewed oil into lakes of black liquid all around. Armenians, Cossacks, Tartars and Persians as well as Europeans flocked there creating complete squalor and a contrast between instant wealth and extreme poverty that was to remain the hallmark of the area until the Russian Revolution and mark deeply the recollections of one of the workers, Joseph Stalin.

It needed, as in the United States, organization to make the rush for 'black gold' into an industry, however. This was provided by the remarkable Nobel brothers from Sweden, Ludwig, Robert and Alfred, the last of whom was the inventor of dynamite and founder of the Nobel prize. It was Robert who first visited Baku in 1873 in search of walnut wood to make riflestocks at the family plant in St Petersburg. He bought oil properties instead and encouraged his brothers to join him. Immensely hard-working and scientifically inclined, the brothers transformed the scene. They imported Pennsylvanian drilling techniques, together with some American drillers, constructed high-quality refineries and built pipelines, ships and rail cars to transport the oil both into Russia and out of it. The export of Russian oil was, almost at the same time, given a huge extra boost by the involvement of the Paris Rothschilds. Originally drawn in by a group of Russian oil merchants in 1883 needing development finance for a railroad of 550 miles from Baku to Batum on the Black Sea, the Rothschilds, who already had oil interests in southern Europe, rapidly invested on their own account not only in the rail connection but also in storage facilities, production leases and the development of an oil export port for the fields at Batum. Russian oil output grew even faster than that of the United States. At the opening of the new century, before Texan oil had made its mark, with a yield of 206,300 barrels a day it actually exceeded the American output of 173,830 barrels per day. Much of the technology used in Russia, particularly for production drilling, was American, but in the construction and design of oil tankers the Nobel brothers and some of the English shipping magnates like Alfred Stuart, proved that Europe was well ahead.

Marcus Samuel and the Early Days of Shell

It was through an edge in shipping transportation that Standard Oil was first challenged by the company which was to become its greatest rival: Shell. Shell was the insignia chosen by the company's founder, Marcus Samuel, the son of an East London Jewish bric-à-brac merchant, in memory of the shells that his family first traded in successfully. His father used to buy unusual shells from sailors returning from the Pacific and turn them into boxes when shell-covered trinkets and frames were becoming all the rage in Victorian London. He had then expanded into trading through the Orient and in 1878 Marcus Samuel and his brother had set up on their own, specializing in trade with Japan. One of the many commodities that they traded was Japanese coal through the Far East and this in turn led them to think of oil.

It was Frederick Lane, one of the partners in their shipping brokers, who first introduced them to Russian oil. Known as 'Shady Lane' because of the number of different deals he always had on the boil, Lane had pioneered the first import of Russian oil into Britain and had become the Rothschilds' agent for selling their kerosene in the country. He had also organized the first export of Russian oil to India. Lane suggested to Marcus Samuel that the brothers might sell Rothschild kerosene through their wide network of agents in Asia. Samuel was certainly interested but the problem was Standard Oil, which was bound to react to any invasion of what it regarded as its territory in China and other Asian markets. So the two men conceived the plan of building tankers which could be accepted as safe enough to pass through the Suez Canal, then closed to oil traffic for fear of fire and accident. At the same time in 1891, Lane negotiated for Samuel a nine-year supply arrangement with the Rothschild company, Bnito.

The coup was a success. Despite intensive political lobbying and legal machinations by Standard, Shell was able to get permission for its tankers to sail through the canal, thus greatly shortening the transportation time and enabling the company to undercut Standard in Asia. There was a minor hitch when it was found that the local customers preferred the kerosene cans that came with the Standard product although they were happy enough to refill them with Shell's oil. Samuel not only overcame the problem by providing container facilities but he also doubled the profit by developing a technique for steam-cleaning the tankers and sending rice and other commodities back on the return voyage.

By the end of the 1890s, concerned that time was running out on his Russian supply contract, Samuel also became involved in production in Borneo through an oil find at Kutei. (The company's old 'nodding donkey' pumps of brass and teak still stand there despite the fact that oil has long since been nationalized.) In 1897 the Shell Transport and

Trading Company was registered. At the dawn of the new century it was already a formidable rival to Standard Oil in the Asian region, owning thirty ships and a whole network of depots in the East. Within a year of the Spindletop gusher Samuel had supplemented his Russian supplies with a long-term contract for Texan crude oil negotiated with Colonel Guffey.

Samuel, however, was not the man to build on these foundations. He was a trader of the Joseph Conrad era who loved the negotiation but not the hard work, who delighted in shipping but not in manufacture. In common with so many other successful businessmen of his time, he was also anxious for a place in society and the English trappings of success – an estate in the country and a title. As Frederick Lane put it bluntly on resigning from the Shell board in 1902: 'Business like this cannot be conducted by an occasional glance in one's spare time, or by some brilliant coup from time to time. It is steady, treadmill work; and unless one is prepared for this, better let it alone altogether.'[1] Samuel had become Lord Mayor in that year, the first Jew elected to the office, and was not prepared for the treadmill.

Henri Deterding

The person who was, was Henri Deterding. A man in the Rockefeller mould, at least in his ambition and his application, Deterding had been born in Amsterdam in 1866, one of six children. His father had died in the Far East when he was 6. Leaving school at 16, he had worked in a bank before voyaging to the Dutch East Indies as a recruit to the Netherlands Trading Society.

Like Rockefeller, Deterding first made headway grappling with figures and book-keeping, learning, as he put it later in his memoirs, 'the supreme advantage of . . . the unravelling habit'.[2] Like Rockefeller, too, his memoirs are full of the hardship of his youthful training and stress the necessity of hard work. Both despised those who wanted it easy, the win-all-lose-all mentality of the driller. But Deterding, a womanizer of considerable energy, obsessed with physical health and possessed with a formidable ego, 'a higher simpleton' as he described himself, was quite different in other ways. He came to oil not through refining but through trading; not, as Rockefeller had, by remorselessly expanding and consolidating from a centre, but by seeing the larger picture and organizing and negotiating to achieve it. The Dutch motto, 'Co-operation gives power', was one of his favourite sayings and he deeply despised price cutting as the weapon of competitive success, calling it 'a refined form of throat cutting'. He wanted order but a negotiated order which did not leave devastation in its wake.

Deterding entered the oil business when recruited as a 30-year-old by J. B. August Kessler to join the Royal Dutch Petroleum Company, which had been founded by royal charter in 1890 to develop the oil discoveries made in Sumatra. Kessler, a man of considerable tenacity and energy, had managed to pipe the oil to a new refinery on the coast. He was now ready to expand through product sales and recruited Deterding as his sales manager. Deterding and Kessler's first task was to beat back an effort by Standard Oil to take them over completely, a move they were able to block with the issue of special shares which could only be purchased by Dutch citizens. Their next job was to go out and compete with Standard in the nearby Asian markets of Southeast Asia, India and China.

The experience taught Deterding several lessons. One was a lifelong distaste for price cutting. It led, he argued, to a constant fluctuation in the market as low prices in one spot had to be cross-subsidized by excessive prices in another, which in turn upset consumption. If you wanted demand to increase, you had to persuade the customer that prices would remain stable. A second lesson was the need for security of supply. In a famous story, Deterding and Kessler were celebrating the arrival of a new tanker delivered to Sumatra when a cable reached them saying that the field was producing salt water instead of oil. Fortunately they were able to negotiate alternative supplies from Russia but ever after Deterding paid particular attention both to production technology and to diversified supply arrangements. The final lesson, if Deterding needed to be taught it at all, was that Royal Dutch would have to seek alliances if it was to withstand Standard Oil.

Shell was the logical answer. It had the outlets, the access to the British Empire and the force of London behind it. Deterding was for it. Kessler, for reasons of preserving his independence, was against. When Kessler died of overwork, travelling by ship back to Holland in 1900, Deterding took over. It was 'Shady' Lane who brought the two companies together. Samuel and Deterding, thirteen years his junior, were hardly made for each other and talks dragged on through 1901 and 1902. Samuel took the view, understandably enough, that Shell being the bigger company hardly needed to give room to Royal Dutch. Shell was, however, overstretched and Samuel was increasingly preoccupied with fulfilling his mayoral aspirations. In 1902 Lane was finally able to bring together the interests of Shell and the Paris Rothschilds with Royal Dutch in a joint company with equal shares, the Asiatic Petroleum Company, formally incorporated in 1903 with Deterding as its managing director.

Deterding's appointment in the key management role has tended to be presented as an act of weakness or foolishness by Samuel, an open invitation for a man of Deterding's character to use the position to aggrandize himself at Shell's expense. This may have been so. Certainly Deterding was not slow to take advantage of Asiatic to use Shell's facilities to increase the market for Royal Dutch oil. Samuel himself, however, saw it as taking care of his Asian business, the better to enable him to develop his European plans. By the time Asiatic Petroleum was launched, Shell had already embarked on a major programme of expansion in Europe based on the Texan oil negotiated by Samuel. A fleet of specially large tankers was being constructed to carry it. Samuel had approached the British Admiralty to get them to convert their fleets from coal to oil and to use his Texan oil. He had also gone into alliance with the Deutsche Bank to import oil into Germany. In Asia, meanwhile, the nationalist Boxer Rebellion in China against foreign investment and harsh competition was causing problems. It made sense for Samuel to reduce competition and ease the trade situation in the East while he pursued his multifarious plans nearer home.

It didn't work out that way. Barely a year after making arrangements to import 10,000 tons of oil a year from Texas, the Spindletop discovery dried up and Samuel had reluctantly agreed to cancel the contract (both Rockefeller and Deterding would have held Guffey and the Mellons to it). Having turned back the advances of Standard Oil, which had offered Samuel £8 million for his company when it seemed strong in December 1901, Shell found itself typically facing a savage price war from Standard through most of its markets in Europe. Samuel wanted to remain British, not least because he had his eye on the Royal Navy contract. But the pressures were mounting. The all-important trials to show off oil's use as a tanker fuel and to display how well Shell, which had developed something of a speciality in bunkering ships around the world, could deliver it ended in disaster in the summer of 1902 when the boilers of HMS *Hannibal*, chosen for the trials, belched out smoke and soot. They had been fitted with the wrong vaporizer burners. It was simple to rectify but the Admiralty was put off both the company and its product for years. Forced to import oil from the Far East and Russia to sustain his marketing effort in Europe, Samuel was impelled to sell six of the tankers originally intended to handle the trans-Atlantic trade to German buyers to cover the dividend. By 1904 he was actively canvassing the British Government for help, both in providing him with new concessions in Burma and in considering Shell for a Royal Navy contract. The Government proved unsympathetic on both counts. Burmese oil production was already in the capable hands of the Burmah Oil Company. The Royal Navy for its part was reluctant to become dependent for its supplies on Shell, a company whose main-owned production was in Dutch territory in the East Indies, half of whose business was carried out in partnership with a Dutch company and French bankers and which also had close ties with the Deutsche Bank together with other German interests.

The Merger with Royal Dutch

Having tried all the alternatives and having been approached by Standard yet again, Samuel was forced to take the obvious route and seek a full merger with Royal Dutch. He travelled to the Hague in April 1906 to propose a 50/50 split. Deterding dismissed the idea. He wanted a 60/40 deal and pointed out bluntly that, in the last few years, his company had consistently outperformed Shell both in terms of growth and dividends. It had consolidated its position in Asia with a series of take-overs of small companies and had successfully negotiated itself, in competition with the Germans, to take a major part in the newly developing Romanian oilfields. Their roles were now reversed. He demanded an answer before Samuel left the Hague, otherwise the deal would be off. Samuel agreed and returned to a general meeting of Shell shareholders to explain what had happened to bring about this 'painful occasion'. Early in 1907 the legal formalities of the new arrangement were settled. Royal Dutch and Shell Transport and Trading became holding companies, the 'Group' as it became known internally, in which the Dutch held a 60 per cent share. The Group in turn controlled three major operating subsidiaries: the Bataafsche Petroleum Maatschappij (BPM), handling production and refining and based in the Hague; the Anglo-Saxon Petroleum Company, based in London and handling transportation; and the Asiatic Petroleum Company, still with its Rothschild interests (they gave them up in the 1920s) handling marketing and also based in London. Deterding, later to become Sir Henri and to own an English country estate, moved his office to London to head the business. Marcus Samuel, already knighted for his help in assisting the rescue of a naval warship in 1898 and later to become Baron and finally Viscount Bearsted, took on more and more the role of the spokesman.

Samuel's greatest disappointment was that the union with Royal Dutch had finally scotched his chances of getting the contract to supply the Royal Navy with fuel oil. 'Never in the annals of British trade has so gross a wrong been done to any company,' he declared to a meeting of shareholders in 1913 when Winston Churchill as First Lord of the Admiralty had publicly rejected the company, 'in classifying Shell as a foreign corporation'. This was a little unfair of Samuel. Although there was at various times – and would continue to be – a distinct strain of anti-Semitism in the Navy's attitude to Samuel, up until 1913 the Admiralty had in fact treated Shell very much on its competitive merits. Part of the company's trouble was that its own production in Borneo, as in so many other regions, was expensive and difficult to refine. Nor did Samuel endear himself to the public when, defending the oil companies against a savage round of petrol price increases in 1912–13 he declared that 'the price of an article is exactly what it will fetch'.[3]

Fuel for the British Navy

The rise of Royal Dutch/Shell and the dominance of the world's oil trade by the two great colossi of Standard Oil and Shell, however, inevitably raised profound concerns among European governments, and the British Government in particular. No country, Winston Churchill told the Commons in June 1914, announcing the Government's momentous decision to take a direct interest in a British oil company, could allow itself to be subject to the 'long steady squeeze by the oil trusts all over the world', or leave itself at the mercy of the two giants who now controlled oil.

In a blistering attack on Shell he claimed:

> It is their policy to acquire control of the sources and means of supply and then to regulate the production and the market price. . . . We have no quarrel with Shell. We have always found them courteous, considerate, ready to oblige, anxious to serve the Admiralty and to promote the interests of the British Navy and the British Empire – at a price. . . . The only difficulty has been price. On that point, of course, we have been treated with the full rigour of the game.

The question of price was very real. Until well into the First World War Britain had been primarily a user of kerosene and, without refining facilities of her own, had imported it from first the United States, then from Russia and the United States and subsequently from Romania and the Far East as well. As elsewhere, the market had been subject to a pattern of violent price wars followed by cartel arrangements to hold the price (Frederick Lane acting as the marriage broker for agreements between the Rothschilds, the Nobels and Shell). These cartel arrangements were then applied to the petrol market as the cost of fuel rose. But in 1913 that market suffered a series of rapid price rises because of various supply difficulties and the Government found its own fuel oil requirements apparently stymied by high price quotations and an unwillingness among the major oil companies to quote for large volumes.

The timing was particularly sensitive, for, under the proselytizing energy of Admiral Sir John Fisher and the political enthusiasm of Winston Churchill, the Government was just on the point of converting its huge fleets from coal to oil. Fisher, the great Navy reformer, had long been convinced of the merits of oil over coal, a conversion which had earned him the nickname of the 'Oil Maniac'. He had favoured Samuel, whose equal enthusiasm for developing markets for fuel oil made him particularly eager to come to an arrangement with the Navy. Indeed, at various times during the period from 1902 to 1911, Samuel had suggested putting a British government director on the Shell board to make it easier to gain the Navy contract, a suggestion which Deterding on the later occasions firmly rejected, pretending to keep his distance from government.

In 1910 Fisher resigned as First Sea Lord but at the end of 1911 he found a new champion for his cause when Churchill moved from the Home Office to the Admiralty. Fisher wrote to him within days, pressing his point about the urgent need to convert to oil. 'Your old women will have a nice time of it,' he snorted, 'when the new American battleships are at sea burning oil alone and a German Motor battleship is cocking a snook at our "Tortoises".'[4] Churchill, always susceptible to new ideas and nationalist sentiment, sought the advice of his officials and found them largely agreed. By the following spring he had taken what he called the 'fateful plunge' and ordered five fully oil-burning battleships.

Having taken that plunge, in typical Churchillian fashion, he then worked backwards to try to sort out the supply problem. A Royal Commission into Fuel and Engines was appointed in June 1912 with Fisher at its head to point it in the right direction. 'You have got to find the oil,' Churchill wrote to him, 'to show how it can be stored cheaply: how it can be purchased regularly and cheaply in peace, and with absolute certainty in war.'

His words expressed perfectly the dilemma for every importing country then and since. In Britain's case the answer eventually turned out to be, most surprisingly, a form of nationalization. In looking round for sources, Fisher and the Admiralty came across the simple problem that, despite the size of the British Empire, it had so far proved to contain few major reserves of its own. There were the Burmese fields, exploited by the Burmah Oil Company, but these were largely committed to the Asian market and particularly to India where they enjoyed tariff protection. Otherwise there were no obviously secure sources, certainly none under British control. Weetman Pearson, later Lord Cowdray, had made important discoveries in Mexico and was selling oil to the Navy. But under the Monroe doctrine, a principle of American foreign policy which opposed foreign interference in the Americas, his fields were in territory regarded as within the ambit of the United States (a fact which the American Government was not slow to point out) and were all too vulnerable, as it proved, to the upheavals within Mexico (see Chapter 6). The United States itself could not be trusted to continue exports in case of war and was, in any case, short of fuel oil itself. The Russian, Romanian and Dutch East Indies' supplies were all in the hands of international companies, which the Government tended to distrust, especially in the case of Standard Oil and Shell. The discoveries in Trinidad and Egypt were relatively small and new and had needed the help of Shell to get going.

Knox D'Arcy and Anglo-Persian

It was with a ready ear, therefore, that the Admiralty heard the pleas for support from the Anglo-Persian Oil Company, which came to it in

1912 for help in developing its huge new finds in southern Persia. The story of the discovery of oil in Persia is one of the legends of the oil industry. It had been known for centuries that petroleum existed there and as in Russia, the right of exploitation in the early years of oil had been given as a virtual monopoly, in this case to Baron Julius de Reuter, the founder of the Reuters wire service. This had come to naught, not least because of the political troubles that saw Persia divided between Russian and British interests in the 'Great Game' of the latter half of the nineteenth century. Against this unsettled background, a new oil exploration concession was sought by William Knox D'Arcy in 1901. D'Arcy, an English-born Australian who had made a fortune out of Queensland gold, was persuaded of Persia's potential by Edouard Cotte, a former secretary to Baron de Reuter. Cotte's brother-in-law had taken part in a recent archaeological expedition to Persia, led by Jacques de Morgan, which had been impressed by the oil seepages it found there.

D'Arcy never visited Persia himself but sent out envoys to negotiate a commission with the support of the British representatives there who were anxious to engender commercial activity as a counter-balance to Russian influence in the north. The terms of the concession were modest enough, although they may not have seemed so generous at the time, given the wildness of much of the country and the lack of government control of the region. D'Arcy was given all rights to find and export petroleum in the country, except in the five northern provinces under Russian influence, for sixty years in exchange for a payment of £20,000 in cash, a further £20,000 in shares and a 16 per cent royalty on the profit.

D'Arcy engaged the services of an ex-Indian public works official, G. B. Reynolds, as a driller and early in 1903 he formed his first company. By the end of that year he was already in financial difficulties and appealed to the Admiralty for help to prevent him having to sell out to 'foreign interests'. Through the good offices of Boverton Redwood, a government adviser on petroleum exploration and an enthusiast for Persian prospects, the Admiralty arranged for Burmah to come to the rescue.

With new financial backing Reynolds now carried out an extensive survey of D'Arcy's concession and decided to concentrate his attention on an area in southern Persia already known to the locals as Maidan-i-Naftun, the 'Plain of Oil'. Drilling was extremely difficult. They were exploring in a remote area, populated by bandits, controlled by local khans and far from any port or point of transport. At one point, when local tribesmen proved awkward, a Royal Indian Marine gunboat had to be sent as a show of strength, a show somewhat undermined by the fact that it immediately got stuck on a sandbank. After three years spent drilling just two wells with a third started, the company showed signs of giving up. 'Cannot Government be moved,' Arnold Wilson, the British official protecting the drilling, wrote desperately to the local consul-general, 'to prevent these faint-hearted merchants masquerading in top hats as pioneers of the Empire, from losing what may be a great asset?'[5] Reynolds, on his own recollection – although this was later denied by one of the Burmah directors – was told by cable to pack up and abandon Persian drilling. He turned a Nelsonian blind eye, saying he would continue until he received formal instructions in writing. Just before he got the letter, he struck oil in a great gusher on 26 May 1908. 'See Psalm 104, verse 15, third sentence,' he cabled back to London. The relevant quotation was: '. . . that he may bring out of the earth oil to make a cheerful countenance'.

A new concession company, the Anglo-Persian Oil Company, was formed in the spring of 1909 to develop the find. It took two years to build the necessary pipe and production facilities to get the oil to the coast at Abadan and it was yet another year before oil exports began. The company refinery was not completed until 1913, by which time the Admiralty, which had taken a new interest in Anglo-Persian and had helped to appoint its chairman, Lord Strathcona, was looking for regular large-volume supplies of fuel oil. By then the company had become severely overstrained financially by the cost of developing its fields and it was also desperately anxious to keep out of the hands of Shell, with whom it had formed a marketing agreement through Asiatic Petroleum in 1912. Once again it appealed to the British Government, playing strongly the 'anything but the "foreign monopolist" Shell' card.

The India Office was cool, saying that it was none of their affair. Admiral Fisher was ambivalent, favouring the idea of secure supplies from Persia but less easily impressed by the 'Shell menace' argument put so forcefully to his commission by Sir Charles Greenway, Anglo-Persian's chief executive. Already close to Samuel, he became positively gushing about Deterding after the latter had given evidence. 'The greatest mistake you will ever have made', Fisher wrote to Churchill, 'will be to quarrel with Deterding. He is Napoleon and Cromwell rolled into one. He is the greatest man I have ever met. Placate him, don't threaten him!'[6] The Foreign Office, however, wanting to maintain a British influence in the area, was in favour of a special arrangement with Anglo-Persian, even though it was unwilling to put any government money into the enterprise. The Admiralty, now under Churchill's control and convinced that Persian oil, being on the coast, was more secure than Russian and Romanian supplies and also deeply suspicious of Shell, was enthusiastic. In July 1913 Churchill was telling the Foreign Office that 'the development of the Anglo-Persian oil supply is indispensable to the solution of the liquid fuel problem', and was hinting to the Commons that government money would be forthcoming. An Admiralty Commission, headed by Vice-Admiral Edmund Slade and including John Cadman, Professor of Mining and Petroleum Technology at Birmingham University, later chairman of Anglo-Persian, went out to the fields and reported back favourably. In the summer of 1914 the Cabinet decided to pump in £2 million in exchange for 51 per cent of the company's equity. Under the Anglo-Persian Act of August 1914, terms were laid down under which the Government, while holding the majority shares and appointing two directors with the power of veto over board decisions, agreed not to interfere in the commercial running of the company. The Admiralty, for its part, got what it had failed to get even Shell and Standard Oil to quote for: a ten-year guaranteed supply of fuel oil under extremely advantageous terms. Marcus Samuel was mortified. Deterding, who received a message from Churchill saying that nothing personal was meant by the remarks he had made in the Commons, remained calm.

Floating to Victory on a Wave of Oil

War did indeed break out later that year and Churchill was proved right. Oil was the vital ingredient. By 1918 the tank had demonstrated its worth as had mechanized transport for the infantry. The Allied armies had over 150,000 trucks in use, 60,000 of them British. The aircraft had come into its own and the Royal Navy had converted almost its entire fleet to fuel oil. 'The Allies,' declared Lord Curzon in a memorable phrase, 'floated to victory on a wave of oil.'

It was on a wave of American rather than Persian oil supplies that Britain had swept to victory, however. The contribution of oil from Abadan never amounted to more then 500,000 gallons during 1918 as compared to well over 1000 million gallons of American oil delivered into Britain, many of the tankers under the severest threat from German submarines. American shipping proved its worth in both world wars. So did the major oil companies. The early years of the First World War, when initial expectations of victory had given way to resentful recognition of the fact that this was to be a long haul, saw considerable criticism within the British Government of the multinational practices of the major oil companies. Shell in particular was singled out for accusations that it continued to make profits out of supplying the enemy through its interests in Romania – then a neutral country – and through its Scandinavian marketing subsidiary. This was certainly true although Deterding, who remained in Britain throughout the war, if anything showed a pronounced prejudice in favour of Britain and was quick to replace the management in Scandinavia when the allegations were made.

None the less Sir Charles Greenway, a cold, devious man with an intense (and mutual) dislike of Deterding and Samuel, who was to be immortalized by Upton Sinclair as 'Old Spats and Monocle' in his novel

Oil, was quick to muddy the waters to gain increasing support from the British Government for his company's ambitions. With the support of the British Custodian of Enemy Property he was able to obtain for Anglo-Persian the sequestered marketing networks of the German importing companies, including, ironically, 'British Petroleum', a distributing company owned by German capital and Russian oil merchants. He also manoeuvred strongly during the war years for a combination of British oil interests into a grand imperial consortium to include Burmah Oil and the Mexican oil interests of Lord Cowdray, who was then offering the trouble-ridden properties to the British Government at half the price he was being tendered by Standard Oil. Greenway's plans were strongly supported by the Foreign Office and at first by the Navy but eventually they came to naught. In part this was because of personality clashes and in part because Anglo-Persian overplayed its hand both on the matter of oil production in the Ottoman Empire and by the violence of its antipathy towards Shell.

Instead, by the end of the war the idea of creating an imperial oil company around Shell, taking in Burmah Oil and possibly a share of Anglo-Persian, gained ground. In 1919 Deterding and Lord Harcourt, chairman of the Petroleum Imperial Policy Committee, set up by the Government in 1918 to work out a post-war oil policy, actually signed an understanding under which the ratio of holdings in the Group was to be reversed to give Britain rather than Holland the 60 per cent stake in exchange for government support for Shell's interests both within the UK market and in seeking oil production sources abroad. It too eventually foundered despite being revived on several subsequent occasions, through a combination of waning government enthusiasm and irritation by Deterding, who expected far more from the British Government than he ever got.

The Post-War Period

This uneasy entwining of commercial and political imperialism coloured the international oil scene for the whole of the inter-war period and well after the Second World War. On the one side were the companies, eager to consolidate their hold on markets and supplies who looked to their governments to support them. Standard Oil (New Jersey), the strongest child of the Standard break-up, emerged from the First World War deeply concerned both at the lack of international oil production to support its marketing operations around the world and indeed about its production position in the United States. Royal Dutch/Shell, which had by this time successfully invaded Standard's own territory in the United States, was equally anxious to participate fully in the Middle Eastern carve-up between the Allies which followed the war and also wanted to gain British support for its recovery of lost production in Romania and Russia. Anglo-Persian, Burmah Oil and

Cowdray's Mexican Eagle all felt they needed the help and umbrella of government to enable them to compete with the two giants of Standard and Shell. On the other side were the governments themselves, anxious to secure oil flows and ensure national interests in the oil industry, but at the same time intensely suspicious of the loyalty of the international companies that were offering the best chance of securing the variety of supply that they were seeking.

'Mr Five Per Cent'

The first testing ground for these competing ambitions proved to be the conquered territories of the Turkish Middle East. Mesopotamia had already been the scene before the First World War of sharp manoeuvring between Anglo-Persian and the British Government to exclude Shell from what was known, on the evidence of the Bible and travellers over the centuries, to be of considerable oil potential around the Tigris and Euphrates rivers in what is now Iraq. Deutsche Bank, then helping to build a railway from Berlin to Baghdad, had gained oil rights to the land on either side. The Sultan, Abdul 'the Damned' as he became known to Hollywood history, had sought the advice of a young Armenian entrepreneur Calouste Gulbenkian, who was convinced that there was oil in the region, and paid him nothing but a compliment for his pains and the survey he made. Upheaval within Turkey and inefficiency in its territories had slowed exploration for oil until the terrible Armenian massacres forced Gulbenkian, a man of enormous charm and possessed with a gift for making himself useful to the right people, to flee Constantinople in 1896 and set himself up to advise Deterding and others on the riches that might lie beneath the land of the two rivers. In 1912 Gulbenkian helped to establish the Turkish Petroleum Company, consisting of the British-owned National Bank of Turkey with a 50 per cent share and Deutsche Bank and Shell with 25 per cent each. Gulbenkian had 15 per cent of the half share held by the National Bank. Anglo-Persian, which by then was trying equally fervently to get concessions in the area, rushed to the British Government to appeal against the move which, it argued, would enable Shell to 'outflank' its Persian oil. The Foreign Office was sympathetic and intervened to impose a new agreement to give Anglo-Persian half the shares in Turkish Petroleum and Shell and Deutsch Bank the remaining half. Deterding and the German Bank each honourably gave up 2·5 per cent to Gulbenkian, bestowing on him forever after the nickname of 'Mr Five Per Cent'.

The San Remo Agreement

When war broke out, Greenway was quick to suggest to Whitehall that Anglo-Persian get the whole of Mesopotamia when Turkey was conquered, a piece of cheek which even his supporters at the Foreign

Office thought was going over the top. In the aftermath of the war the whole issue became entangled not only with the question of the creation of a British imperial oil company around Shell or Anglo-Persian but also with Franco–British relations in the carve-up of the old Ottoman Empire. Gulbenkian was busy lobbying the French Government on Shell's behalf while Greenway was busy lobbying London and Paris. There was confusion at the top where Lloyd George, who loathed Samuel and disliked the oil companies, was negotiating quite separate agreements with Clemenceau, the French Prime Minister. Eventually the San Remo Agreement was signed on 24 April 1920. As part of an overall political settlement which gave Britain control of what is now Iraq and France control of what is now Syria under a League of Nations mandate, the Turkish Petroleum Company was given the oil rights. French interests were given 20 per cent of the company, Shell and Anglo-Persian were to have 70 per cent between them with Gulbenkian's 5 per cent taken out of that, and local Iraqi interests were to have 10 per cent.

San Remo was to cause political problems between Britain and France for half a century. It was also to raise endless difficulties in the oil sphere. The French chose to use their share as the basis for founding an oil company of their own in 1923, the Syndicat National d'Étude des Pétroles, instead of using Shell's French subsidiary as Deterding had hoped. The company was later renamed the Compagnie Française des Pétroles (CFP) with a French government stake of 35 per cent. Deterding, having been initially keen on the idea of turning Shell into a British-dominated imperial company, now grew doubtful of its advantages to him, even more so when the Americans expressed strong resentment at the agreement, from which they were excluded. At the time the United States was going through one of its regular periods of geological pessimism. Government surveys suggested that oil would run out within a decade. Oil companies were therefore anxious to get new sources abroad and eagerly sought, and obtained, State Department support for gaining entry to the Middle East, Eastern Europe and other territories. The American Government, with some justice, demanded that the Europeans follow an 'open-door' policy towards allowing American interests into oil concessions, just as the United States had done to Shell, which was now actually a larger oil producer there than Standard Oil (New Jersey).

The situation was made more delicate for Shell by the fact that it had recently taken over Cowdray's Mexican production. Cowdray, who respected Deterding but despised Samuel (indeed he demanded that Samuel be removed from Shell as part of the price of the deal), had grown fed up both with Anglo-Persian and the British Government and decided to proceed with his own commercial interests. With its Mexican reserves and with new drilling concessions in Venezuela,

Shell needed to sell into the North American markets and was none too happy when the American Government under Woodrow Wilson began to retaliate against European companies by refusing to allow them to bid for concessions on federal or Indian territory. Herbert Hoover, then Secretary of Commerce, went much further, suggesting that the American companies form themselves into a syndicate to face down Shell and foreign interests on the international scene. A government which had once broken up Standard Oil now brought together seven of the biggest companies, including three of the former Standard companies – Standard Oil of New Jersey, Standard Oil of New York and Atlantic Refining – to pursue American interests in the Middle East.

The Red Line Agreement

Whatever the politics of oil, the politics of nations could not allow this strain between the Allies, still less with one whose efforts had saved the British and French in war. With the encouragement of both

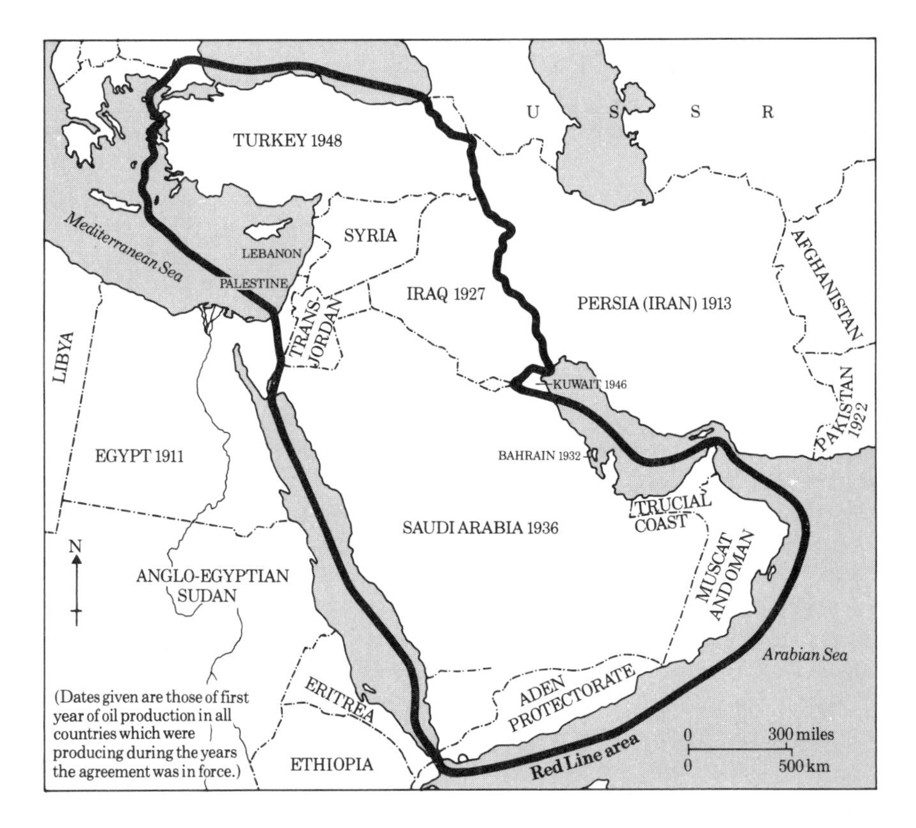

The Red Line Agreement, 1928, drawn up by Gulbenkian to mark the boundaries of the Ottoman Empire. Members of the Turkish Petroleum Company agreed not to compete within the area. Kuwait was excluded (opening a back door to American companies).

governments, in 1922 Sir Charles Greenway invited the new syndicate to join Turkish Petroleum. Negotiations dragged on for years as the overall political situation was sorted out. In July 1928 settlement was finally reached. It was agreed that the American syndicate of five (two members had lost interest), Anglo-Persian, CFP and Shell each took 23·75 per cent, leaving 5 per cent for Gulbenkian. The Iraqi interests, that had once been promised 10 per cent, were by this time shamefully cut out altogether. The agreement was to cover all the territories of the Ottoman Empire. Taking a red pencil, in the meeting Gulbenkian drew a line around the Middle East from Turkey at the top to Saudi Arabia at the bottom commenting: 'That was the Ottoman Empire which I knew in 1914. And I ought to know. I was born in it, lived in it, served it. If anyone knows better, carry on.'[7] Persia was excluded, of course, and so was Kuwait, to the later regret of many of the participants.

The Red Line Agreement served as a forerunner of later carve-ups of both the Middle East and other countries between the key international oil companies. It also marked the dawning of inter-oil co-operation. The nationalistic rivalry of the immediate post-war period was being replaced by a growing awareness of the fact that the multinational oil companies were genuinely international. They had more in common among themselves than with their governments.

Inter-Oil Co-operation

With new discoveries popping up everywhere from Venezuela (in 1922) and Texas to Iraq and the Middle East, the fear of shortage now gave way to the fear of glut. More clearly than anyone else, Deterding at Shell saw that glut demanded co-operation if it was not to lead to disastrous bouts of price cutting on the old Rockefeller scale. He had expressed the view early on during a visit to New York in 1907 to meet John Archbold, then head of the still unbroken Standard empire. Deterding propounded the view that he also gave to a trade paper at the time. He very much wanted to:

> . . . realize our ideal, which is: all points of the globe the same reasonable price for all the products of oil, so that the consumer, the manufacturer, and the merchant may find it worthwhile. Today, with the strangling system of the Americans, certain people pay much too dearly for oil, whilst in other countries oil is thrown into the water. Because of this, many producers are at a standstill, many refineries are ruined, while subsidiary industries cannot make progress.[8]

Archbold, a man of the old school, was unconvinced. By the 1920s, however, Standard was in a very different position. It needed co-operation to sustain its worldwide markets and it was headed by a man of like views to Deterding, Walter Teagle, who had originally introduced Deterding to Archbold. 'The world', Deterding wrote to Teagle in

1925, 'is now suffering from over-production, over-refining, over-transporting and – last but not least – over-retailing.'[9]

Merger was one avenue but it was one that had political problems attached as the Americans reacted against the spectre of a huge British combine and as nationalism in such countries as Venezuela, Romania, Mexico and even Persia led to active discrimination against companies thought to be too closely tied with imperial governments. The other route was co-operation and cartelization. Teagle, whom Deterding described on their first meeting as 'one of the most straightforward men of the Standard Oil Company and a gentleman in every respect', was a frequent guest at the grouse shoots on Deterding's estate. Likewise the old quarrels between Shell and Anglo-Persian had been buried and their UK marketing interests eventually combined, with those of Shell to become Shell-Mex and BP in 1931.

The Achnacarry Agreement

In August 1928 the three strongest men in the international oil industry – Henri Deterding, Walter Teagle, and Sir John Cadman (now head of Anglo-Persian) – came together at the Cameron castle of Achnacarry in Scotland. They were to be joined by others, including one of the Mellons from Gulf. Ostensibly the reason for this constellation was for fishing and grouse shooting. But, as Teagle put it, 'the problem of the world's petroleum industry naturally came in for a great deal of discussion'.[10]

The 'discussion' lasted two weeks and ended in a secret agreement to divide up the world oil market. Under the agreement, known formally as the 'Pool Association' but more popularly called the 'As Is Agreement', the companies accepted that each would share production and markets on the basis of the balance then prevailing and that new facilities were to be added only 'as are necessary to supply the public with its increased requirements of petroleum products in the most efficient manner'. What was more, prices were to be maintained by fixing them on the basis of the higher American costs. Everywhere in the world oil prices were to be determined by the price at the Gulf of Mexico plus a notional standard freight charge for shipping oil from the United States to the overseas market regardless of whether the oil had actually come from the United States or was, for example, merely shipped from Persia to India. This 'Gulf Plus' Agreement ensured protection for American oil and the 'phantom freight' charges made large profits for the companies. Competition was to be stilled, profits protected and the benefits of cheaper supply or greater efficiency were to be kept by the company rather than handed on to the customer.[11] 'Oil profits,' as Lloyd George had commented some years before, 'generally seem to find their way by some invisible pipeline to private pockets.'[12]

The Achnacarry Agreement, eventually signed by Gulf, Standard Oil of California, Texaco and Mobil, was more a statement of intent than an absolute rule of conduct. Despite the formation of the Export Petroleum Association in the United States to combine its major oil shippers and the efforts, for conservation reasons, of the American Petroleum Institute to control American production, it never really worked there and was soon put under enormous strain by the Depression of the early thirties which saw all the major companies suffering substantial losses and having to compete for their market share. In particular markets, however, such as the UK, Scandinavia and Germany, as well as India and Asia, the subsidiaries of the companies concerned met continuously to organize the market, share out the major contracts and avoid undercutting each other, a practice that continued until well after the Second World War.

The Second World War

The Second World War even more than the First showed the crucial importance of oil. The strategic thinking of both Germany and Japan was perforce driven partly by the need to secure supplies which they lacked themselves. Hence Germany's drive for the Romanian oilfields and Japan's invasion of Indonesia. Both poured research, with surprising effectiveness, into the development of synthetic oils from coal and other sources. Once again the Germans attempted (and at first successfully) to strangle Britain's oil supplies through submarine attacks on the tankers and with their attacks on American coastal shipping, even went so far as to force rationing in the northeastern states until the construction of a pipeline from the southwest to the northeast solved the problem. The Allies' push into France and on into Belgium and Germany after D-Day in 1944 was made possible by one of the more remarkable engineering achievements of the war, the laying of a flexible oil pipe, Pluto, from the Isle of Wight to Cherbourg to provide a constant flow of fuel for tanks and transport.

For the oil companies, the war brought a revival of old sores about loyalties. Deterding, who had become fiercely anti-Communist after his second marriage to a white Russian refugee in the 1920s, became increasingly pro-German on his third marriage to a German girl in the 1930s. Autocratic by nature, he hated anything to do with socialism and saw, as did many others, the first rise of Mussolini and Hitler as purifying forces in a decadent world. 'If I were dictator of the world, I would shoot all idlers at sight', he wrote in his memoirs, instructing the printer to put this threat in bold type. By this time he was already becoming something of an embarrassment to his colleagues at Shell with his volatile interventions. In 1937 he was eased out of the company to retire to a German estate. When he died in 1939, six months before the outbreak of war, there were wreaths from both

Hitler and Goering at his funeral and a conspiracy of shamed silence in the company that lasted for a generation after. The man who had so impressed every member of the British and the French governments in the 1914–18 war with his genuine loyalty to the Allied cause, ended his life a virtual exile in the country of the enemy, with his name and his bust exorcized for a long time from the company's offices.

'I don't believe personally that he was a convinced Nazi,' suggests John Loudon, chairman of Shell for much of the post-war period, 'I think he was the type of man who had had such a success in life – and this happens quite frequently – they believe that they know the answers to all the problems. He thought that the way Hitler was going about putting Germany back on the map was a worthwhile effort . . . after the war, especially in the Netherlands, which had been occupied by the Germans, they hadn't forgotten the fact that he had been all for the Nazis before he died. But it was a passing phase. Within a few years his portrait was again hanging in the boardroom and I think wiser minds had to admit that he was the cornerstone in building up this international oil group.'

War brought problems, too, for New Jersey Standard and for Texaco, one of the new giants from Texas and an eventual signatory to the Achnacarry Agreement. In 1937 Walter Teagle was retiring, having taken New Jersey Standard (known as Esso since 1926) from the lopsided survivor of the break-up of Standard Oil to a company to equal Shell in size and even exceed it in production. When war broke out between the United States and Germany in 1941, he and his successor, Bill Farish, were to find themselves under investigation by the Justice Department for research agreements with the German chemical giant, I.G. Farben. Under an agreement made fifteen years before, Standard Oil had handed over its research and patents on the lead used in petrol and aviation fuel to improve performance in exchange for getting the benefits of Farben's research into synthetic rubber, a key product in the war. Esso came under violent jingoistic attack in the press for having sold the country's technical birthright to the enemy and deliberately stifling research into rubber at home all for the sake of a monopolistic trade agreement. The attacks were greatly exaggerated and the anti-trust actions ended in a fine. None the less they caused immense damage to Esso's reputation, so much so that Rockefeller's son was brought in to disavow them. They were to be fanned again just after the war when a congressional subcommittee sent out investigators to Germany who brought out strong evidence from I.G. Farben that they had got the better of the bargain and that Esso's technology had been of immense use to them in the war.

Farben's defence of why it had reached an agreement at all with Standard was illuminating in that it showed how the major oil companies appeared to the countries in which they operated which did not have strong oil companies of their own. As a Farben document explained:

> The closing of an agreement with Standard was necessary for technical, commercial and financial reasons: *technically*, because the specialised experience which was available only in a large-scale industry was necessary to the future development of our process, and no such industry existed in Germany; *commercially*, because in the absence of State economic control in Germany at the time, I.G. [Farben] had to avoid a competitive struggle with the great oil powers, who always sold the best gasoline at the lowest price in contested markets; *financially* because I.G., which had already spent extraordinarily large sums for the development of the process, had to seek financial relief in order to be able to continue development in other new technical fields.[13]

The charges against Texaco were more straightforward. It was accused of having aided Franco during the Spanish Civil War, despite American neutrality, and of supplying Germany with oil even after the British had embargoed shipments there, a policy which the United States tacitly supported even before it entered the war. Texaco even entertained Germany envoys sent to the United States to persuade Americans not to support the British. None of this was technically in breach of American law but it was felt at the time to be against the spirit of American policy and certainly against the interests of her British ally, where Texaco also had large interests.

Much was made after of these incidents to suggest the uncontrollability of the international companies when their interests conflicted with national policies. The same themes were to recur in the British trade sanctions against Rhodesia in the 1960s and in the Arab embargoes against the United States during the sixties and early seventies. For the time being, however, they seemed, and genuinely were, isolated incidents. 'The whole question of control', wrote Sir Robert Waley Cohen, a Shell director in 1923 when talks about a new British mega-company were at their height, 'is very largely nonsense. It is a matter of sentiment, but if by transferring control to the Hottentots we could increase our security and our dividends I don't believe any of us would hesitate for long.'[14]

It was that commercial drive that was to worry governments of importing countries in succeeding years and to drive both producers and consumers to set up state oil companies to outflank the power of the major oil multinationals. As Churchill had observed, they were always ready to serve countries – 'at a price'. The struggle for the future was that price, not the 'absolute security in war' that Churchill had looked for in the First World War. On that score, the oil companies emerged from the Second World War stronger than ever, dominating the international trade in oil and ready to take full advantage of the boom in demand that was to follow the war's end.

3

Sisters under Siege

If the United States had dominated the first half century of the oil industry, the international oil companies dominated the second. Dubbed *le sette sorelle*, or 'Seven Sisters' by Enrico Mattei of the Italian state oil company, Ente Nazionali Idrocarburi (ENI), during the generation following the end of the Second World War seven companies – Esso, Mobil, Chevron, Gulf, Texaco, BP and Shell, with CFP of France sometimes added as an honorary eighth – sat astride the world's oil trade. They owned most of the reserves both in the United States and outside it in the Middle East, Venezuela, Indonesia and the other major oil-producing countries. They decided how much oil to take, from where, at what price and to which market. They shipped it in their tanker fleets, processed it in their refineries and sold it under their brand names in virtually every major world market.

By the mid-1950s, when demand for oil began to 'take off' in the industrialized nations, the seven majors controlled nearly 90 per cent of oil production and 90 per cent of oil sales in the 'free world' outside the United States. Even within it they owned roughly half the reserves and accounted for nearly as much of the home sales. Fifteen years later, in 1970, they were still responsible for 61 per cent of crude-oil production outside the Communist countries and 56 per cent of product sales, for all that nationalization and new competition had done to nibble at the edges of their control. No other industry has ever been organized for so long by such a tight group of international companies with such common interests, let alone an industry of as much importance to the world economy as oil.

Despite the tendency of Mattei and others to lump them together, the companies were not all of a kind. Each had, and still has, for the six that are left, their own personality and history. Esso (now Exxon) together with Shell – the great rivals since the beginning – were always the biggest of the companies. They still are with the most widely dispersed reserves and the most firmly based markets, their management trained in an international, corporate ethos that sets them apart from the other sisters.

Exxon

Exxon has always regarded itself as inheriting the mantle of Rockefeller. Standard Oil (New Jersey), or Jersey Standard as it became known, remained the original holding company after the trust was dissolved in 1911 with John Archbold, former head of the Standard empire, as its first president. The hard-drinking son of a Baptist preacher, Archbold switched sides after leading the Producers' Union in their fight against Rockefeller's freight rates war in 1872 and had even signed the pledge on John D.'s persuasion. In the years that followed he became one of the toughest exponents of Standard's policy of consolidation through take-over. A frequent witness before the steady stream of congressional investigations into his company's practices, he was once asked what it was he did as a Standard Oil director. 'I am a clamourer for dividends,' he replied. 'That is the only function I have in connection with the Standard Oil Company.'[1] He himself became notorious when William Randolph Hearst, as part of his own independent political campaign during the 1908 presidential election against Rockefeller and Standard Oil obtained and printed a file of letters from Archbold to leading political figures revealing that several senators and congressmen were retained on Standard's secret payroll.

Archbold in turn had been succeeded by Walter Teagle, the real creator of Jersey Standard through the twenties and thirties. Teagle was the son of a Pennsylvania oilman who had been one of the many to sell out to Rockefeller. President of the company for twenty-one years, taking three months off each year to go shooting and fishing, Teagle was typical of the oil families which dominated the corporations in their early years – the Everests at Mobil, the Frasers at Anglo-Iranian and the Loudons at Shell. It was Teagle who had finally switched the company's policy from one of confrontation to accommodation at Achnacarry and it was Teagle too who drove it to expand abroad.

An early effort to counterbalance Shell's influence in Europe by buying out the Nobels' oil interests in Russia after the First World War ended ignominiously with Standard paying $11·5 million for properties which it could not hope to operate in view of the revolution that had swept the country and taken over Baku. Shell itself had made the same mistake. Soon after, however, the company proved the main beneficiary of the American Government's successful campaign to include American interests in the Iraq Petroleum Company and gain a part in the exploration and development of oil in the neighbouring states of the Gulf both before and after the Second World War. It also obtained major interests in Venezuela, Latin America and Indonesia making it the 'biggest' in virtually every department of the energy business from oil reserves to gasoline markets, from natural gas to coal and even uranium in the seventies.

The one ambition that constantly eluded Jersey Standard was the Standard brand name. Esso, derived from the initials S.O. for Standard

Oil, was introduced in 1926 and was almost immediately challenged in states such as Ohio and in the Midwest where other Standard companies were operating. An endless series of legal cases culminated in a Supreme Court decision in 1969 denying the company's right to the brand name. The corporation, by now very much run by committee, established a task force to consider a new corporate sign. Nearly two years later the computer, having sorted through 10,000 names, came up with the name Exxon. It cost, it is estimated, more than $100 million to change all the name signs on 25,000 petrol stations, 300 million sales slips and 11 million gasoline credit cards, never mind the plaques on some 18,000 buildings, storage tanks, refineries and other installations.[2] Esso remained its brand name outside the United States.

Shell

Shell, in contrast, had actually pre-empted Esso in selling its products through forty-eight mainland American states and it had been much more successful in establishing its symbol, the yellow shell of Marcus Samuel's youth, on its filling stations from Indonesia to Indiana. Traditionally regarded as stronger on marketing than on finding oil, it attempted to rectify this with the purchase of Wolverine Petroleum in the 1920s and of a controlling interest in American Petroleum in the early 1950s. It also sought to increase its production through exploration concessions in California, Mexico (before nationalization in 1938) and, most successfully, in Venezuela. What Iran was to BP, Venezuela became for Shell. Until the company started to develop Nigeria in the 1960s, it provided its biggest single source of own-operated crude oil as well as being the training ground for most of its top management. The experience of Deterding gave Shell a profound suspicion of central dominance. A unique system of rule by a group of managing directors, balanced three-to-two in favour of the Dutch with the chairmanship rotating between the two nationals, was set up to prevent one-man rule by a Deterding ever recurring. Management was devolved from the centre to the companies in each area and the Group deliberately set out to develop a breed of managers who were totally international in outlook.

BP

BP, unlike Shell, remained a very British company, an oil-rich group in search of outlets, for long imbued with a semi-colonial spirit born of its years in the Middle East, run partly by dour Scottish accountants, a tail constantly wagging the dog, in this case the British Government, its principal shareholder. The decade before the Second World War had seen the company strengthen its position in the Middle East with its major stake in the Iraq Petroleum Company and associated enterprises, as sole developer of Iran and as a partner in the exploitation of concessions in Kuwait. In the post-war years it sought to build on its crude wealth by buying into the main markets, sometimes in co-operation with the other majors, sometimes in competition. It was not until the end of the 1960s, however, that the giant North Slope discovery in Alaska finally gave BP the lever to open the door of the United States through the negotiations of a major interest in the Standard Oil Company of Ohio (Sohio), an oil-short marketing company in the American Midwest.

The four remaining sisters, all American, had become international majors largely through the marketing of indigenous oil in the heady days of soaring American exports in the 1920s. Two, Chevron and Mobil, were the step-children of the Rockefeller empire. The other two, Texaco and Gulf, were the sons of the Spindletop gushers.

Mobil

Long regarded as something of a junior partner in the sisterhood, Mobil, which was to become one of the most outspoken of all the oil companies in the seventies under the remarkable duet of its suave chairman, Rawleigh Warner and its president, the fast-talking and skilful-dealing William P. Tavoulareas, was in fact the product of a merger between two Standard companies. One was Standard Oil of New York, the administrative arm of the Rockefeller empire which held a near monopoly of gasoline retailing in the surrounding area under the brand name Socony. The other, the Vacuum Oil Company, was a refining concern founded by Matthew Ewing and Hiram Bond Everest to develop a vacuum system for processing oil. It owed its early success not to the automobile but to the horse-drawn carriage. The manufacturing process invented by Ewing produced kerosene at no better cost than rival methods but it did make an oily residue which Everest found was particularly good for lubricating carriage harnesses. Vacuum Harness Oils were a considerable success and the company concentrated on marketing its own range of lubricants including a cylinder oil for large machinery. It was profitable enough to tempt Rockefeller who took a controlling interest in it in 1879.

Socony and Vacuum had always been 'crude short'. When their traditional supply arrangements with other companies in the Standard empire ceased with the 1911 break-up, Vacuum chose to concentrate on its manufacturing base by building up refinery capacity. Socony pushed for its own production, buying into in 1918 a Texan oil-producing company, Magnolia Petroleum, headed by J. S. Cullinan, one of the founders of the rival Texaco company and, in 1926, the General Petroleum Corporation of California. In 1930 it added a midwestern arm through the White Eagle Oil and Refining Company.

Socony's main presence abroad was always in the Far East, especially in China where it had launched a powerful sales campaign to promote kerosene by designing and giving out with the fuel a cheap oil lamp, called the *Mei Foo* from the Chinese symbols for Socony, which meant 'beautiful confidence'. Socony made its two most crucial moves towards expansion in the thirties. In 1933 it joined forces with Esso in Asia to combine Esso's Indonesian production and refining with its own Asian outlets in a joint company, Stanvac, a combine finally dissolved by American anti-trust action in 1962. And in 1931 it merged with Vacuum Oil to form Socony Vacuum. This was later changed to Socony Mobil and finally just Mobil, the brand name of their petrol, in 1966.

Socal

While Mobil never quite managed to find the oil it wanted to feed its markets either within the United States or outside it, the third Standard company, Standard Oil of California, was always in the opposite position. Socal, as its full name gets shortened to (or Chevron as it is also called, after its petrol brand name), had been the victim of one of Rockefeller's most savage price wars. Originally called the Pacific Coast Oil Company, it was founded in 1879 by a wandering Pennsylvanian oilman named Demetrius Schofield. The company discovered oil near Los Angeles but also found itself up against Standard Oil, then shipping oil into California from the Gulf of Mexico right round South America. Pacific Oil had the cost advantage, Standard had the resources. Over several years prices were selectively cut to 17·5 cents per gallon as against an average elsewhere of at least 25 cents. Finally Pacific gave up, selling out to the enemy in 1900. When Standard Oil was broken up, Socal took with it some of the company's biggest reserves. By the end of the First World War it was producing a quarter of the total American production. By the beginning of the Second, it was on the threshold of achieving the same target internationally as the concessionary company in Saudi Arabia. It applied this advantage with speed in negotiating a joint agreement with Texaco to merge production, refining and marketing facilities throughout the Eastern hemisphere east of Suez, in the Caltex partnership in 1936.

Isolated from the other major oil companies in San Francisco, dependent on the single relationship with the Saudi Arabian Government for its international operations and obsessed with secrecy, Socal was always regarded as the most conservative of the Seven Sisters. It was not until well into the sixties that it even allowed women secretaries on to its top executive floor. Its partner in Caltex, Texaco, long had a similar reputation, only in its case, this extended to tight-fistedness as well as secrecy.

Texaco

Texaco was founded, as mentioned, by a former Standard Oil employee, J. S. Cullinan, who made good at Spindletop when Standard had thought the chances of finding oil there were minuscule. Cullinan, who went into partnership with a New York banker, Arnold Schlaet, to set up the Texas Fuel Company in 1902 had a genius for finding oil. He first discovered it near Spindletop and built a 20-mile pipeline to the coast. When Spindletop ran dry, he found it again in Sour Lake, Texas and in the Indian lands of Oklahoma and Louisiana. By 1904 he was producing 5 per cent of the total American crude-oil output. He sold the oil for lighting, to fire the boilers of the southern sugar plantations and he pioneered its use in railroad transport. The company developed a considerable sales network in Europe and, under the insignia of the Lone Star, was one of the earliest and most aggressive gasoline marketers right across the United States. Like Deterding, Cullinan did not suffer idlers gladly. On one occasion, before starting to extinguish a production fire, he demanded the right to shoot any helper who did not obey orders. Nor did he easily forget hurts. When, like so many of his fellow pioneers, he was forced out of the company by the money men in an internal dispute in 1913, he immediately moved across the road to set up a new company, flying the skull-and-crossbones from Houston's Petroleum Building 'as a warning to privilege and oppression'.[3] The leadership of the company moved to the art deco stylishness of the Chrysler Building in New York and the post-war imperiousness of Augustus C. ('Gus') Long, the son-in-law of Cullinan's lawyer, a man renowned as much for his stinginess in cost control as his conservatism in taking business risks.

Keeping to the business it knew best, that of oil and oil marketing in particular, Texaco expanded in the post-war period through acquisition: of Regent Oil in Britain and the Trinidad Oil Company in 1956, the Paragon group in the United States in 1959, the White Fuel Corporation in Boston in 1962, the Superior Oil Company in Venezuela in 1964 and Deutsche Erdöl of West Germany in 1966. It was an acquisition, that of Getty Oil in 1984, that was to bring the company to the greatest crisis of its history when the alternative bidder, Pennzoil, sued it for wrongful gain the following year (see Chapter 5).

Gulf

While Texaco and Socal in their international operations became primarily sellers of their Saudi Arabian crude-oil wealth, the other beneficiary of Spindletop, Gulf, became equally dependent on Kuwait, where it formed a partnership with Anglo-Persian in 1934. Gulf in its early years was very much the creature of the Mellon banking family, who pushed out the expansive Colonel Guffey in 1907 ('I was throwed out,' Guffey stated forcefully if somewhat inelegantly later) and used their resources to build up the company as a major refiner and gasoline marketer in the southwest. Lacking the extensive refining and marketing facilities of their fellow majors in the international arena, they long remained more of a producer and shipper of oil to others, particularly Japan and the Far East, than a fully integrated oil company. Accustomed to the personal relationships and the corruption of concession arrangements around the world, they found themselves the subject of a dramatic series of revelations of bribery when a periodic wave of business puritanism swept the American political scene after the Watergate fiasco in the 1970s. Gulf officers were shown not only to have given in to a blatant demand for $4 million in bribes to allow them to build a refinery in South Korea but also to have been regularly smuggling into the United States packages of cash with which to pay legislators and senior politicians in Washington and elsewhere. Altogether investigations revealed that Gulf had given out some $12 million in illegal payments through the previous decade.

In 1984 Gulf Oil, harried and pursued by a hostile take-over bid from T. Boone Pickens and a group of speculators, sought refuge in a $13-billion merger with Socal – the biggest merger in American corporate history and the first death in the once-impregnable and undefeatable ranks of the seven.

The Search for Integration

To those on the inside, these Seven Sisters were a fairly motley group of companies, with little love for each other and quite different *esprit de corps*. To those on the outside, they were a formidable, if not impregnable, force, controlling virtually all of the Middle East reserves in the 1950s and 1960s, an era dominated by Middle East output, as well as most of the markets. Both for the companies and for the industry as a whole the search was on for integration, an imperative to match up the two ends of the business, the 'upstream' of oil production and the 'downstream' of gasoline and fuel marketing. Those like Shell and Mobil with markets but without adequate oil sources pushed constantly to gain access to the low-cost and high-volume reserves of the Middle East or into newer areas that might replace them such as Libya and Nigeria. Those with oil but lacking historic markets like BP and Texaco pushed in the opposite direction, buying into marketing companies in Europe, as Texaco did with Regent, a British company, in 1956 and Deutsche Erdöl in West Germany ten years later, or setting up joint marketing facilities with other majors, such as Caltex and Stanvac in Asia or Shell-Mex and BP in Britain.

From the point of view of the industry, the dominance of the trade by these major integrated companies had the powerful attraction of enabling the international movement of oil to expand at the enormous pace required by the market. The drive of these companies was to lower costs through larger and more flexible facilities. The average size of tankers, of refineries, of filling stations and of pipelines doubled and then trebled through the 1950s and 60s. The huge profits made by the companies from their low-cost oil production enabled the industry, astonishingly for a business of its size and scale, to be nearly self-financing. Co-operation between the companies allowed huge joint facilities such as the Tapline pipe to take oil from Saudi Arabia to the Mediterranean to be built in 1949. In addition, their geographical balance gave them the adaptability to adjust supply and the sources of oil to meet the final requirements of a consumer whose demand was more than doubling each decade.

Whatever the criticisms, a number of economists and observers, as well as government officials and the oil companies themselves, maintained that the unique nature of the oil business and its unprecedented growth required a degree of power concentration that was not needed by other industries. It was a matter of logistics, of matching supply and demand through a complex series of transport and processing operations that required large sums of capital, an assurance of through-put and sales and long-term planning management that could not be compared to relatively simple businesses such as automobiles or steel.

It was not an argument that particularly appealed to other economists such as Professor M. A. Adelman of the Massachusetts Institute of Technology, who claimed that oil was no different from other commodities or businesses and that the dominance of the major oil companies and lack of competiton effectively ensured an inflated price structure. It was not a situation that made either consuming or producing governments entirely happy either in the immediate post-war period or later.[4]

Led by their military procurement agencies, both the British and American governments strongly contested the majors' system of setting prices on the basis of the Gulf of Mexico in the years immediately after the Second World War. Details of the Achnacarry Gulf Plus Agreement were eventually prised out of the companies by the Federal Trade Commission and published in the Senate Small Business Committee's, report, *The International Petroleum Cartel* in 1952. Middle East oil, clearly much cheaper than American oil and much nearer the main consuming areas of Europe and Asia, was being penalized by the agreement which was any way becoming progressively less relevant as direct American oil exports to Europe began to dry up after the war.

First the American and British navies and then the Organization for Economic Co-operation and Development (OECD), set up to help organize the Marshall Plan in Europe, objected to the oil prices which were demanded from them, finally getting a separate pricing, or 'posting' as it was called, for crude oils in the Persian Gulf. Both the Senate and the Trade Commission tried to go further and force the issue of anti-trust against the companies. They were effectively prevented from doing so, however. The very reason that made the Middle East now so important – the drying-up of American exports – also made the American Government more concerned with the Middle East both as a potential threat and as an important source of imports. Whatever the drawbacks to a system of trade dominated by a few major oil companies in terms of prices, it did at least ensure Anglo-Saxon control, a point that was equally valid in London. So whether it was the State Department in Washington or the Foreign Office in London, their resistance always curbed the attacks of other government departments on the status quo.

For other countries, such an Anglo-Saxon monopoly was, of course, far less satisfactory. The Germans, Austrians, Italians and Scandinavians – and for that matter the newly developing countries such as India – consistently sought ways of promoting their own companies, whether state or private. The oil-exporting countries – some like Venezuela more than others like the smaller Trucial States – toyed with the idea of developing national corporations that would handle their resources. The difficulty was that the consumers couldn't get at the cheap oil resources which might break the majors' near monopoly of Middle Eastern production and the producers couldn't get a secure enough access to the final markets to enable them to do away with the major oil companies.

Enrico Mattei

It was Signor Enrico Mattei who attempted to burst through this stalemate and bring the producer and consumer in more direct contact. Part demagogue, part magnate and part visionary, Mattei electrified a reviving but still humiliated post-war Italy with the idea that it could take on the Anglo-American cartel and play a pivotal role in a new world of Arab nationalism. In a series of speeches, interviews and articles, he singled out the oil companies, the Seven Sisters as he called them, half in awe, half in bitter resentment, as the enemy and the State as the expression of popular will that would sweep them away.

Whether he really expected to engulf the world with a new tide of state-to-state oil trading, with Italy riding on the crest of the nationalist wave, has never been certain. As Anthony Sampson remarks in his book, *The Seven Sisters*, Mattei 'was fascinated, perhaps too fascinated, by the mystique of the sisters and their power: and it was not quite clear whether he wanted to beat them, or join them'. This ambivalence of attitude was to be repeated in the case of many of the heads of the state oil companies that mushroomed among the producing and the consuming oil countries such as BNOC in Britain (in 1976), Statoil in Norway (in 1972), Elf-ERAP in France (1966), Pemex in Mexico (1938), NIOC in Iran (1952), Pertamina in Indonesia (1968) and Petromin in Saudi Arabia (1962). Appointed in part on political grounds, imbued with a powerful and ambitious energy to make their companies grow from small national beginnings into international stars and boosted on their way by oil revenues far greater than their industrial sister state corporations, the heads of these companies have always been tempted by the grander stage and a corporate expansionism that has tended at times to outrun political control.

Mattei, once an oil official of Mussolini's and later a partisan leader during the war, a man close to the ruling Christian Democrats in the post-war years, was appointed as head of the newly established state oil company, ENI, in 1953. His aim was to make Italy self-sufficient in oil, partly in opposition to the international oil companies and partly in competition with the Italian electric power industry. Early on it looked as if he might bring off the feat in spectacular style when he discovered substantial gas reserves in the Po Valley and signs of oil within Italy itself. Unluckily the gas proved to be limited in volume and the oil never amounted to much more than a smidgen. So Mattei set himself the task of gaining cheap and secure supplies from abroad. Refused entry to the new consortia being formed to participate in Iranian production following the upheavals there and the demise of Mossadeq in 1953, he decided to do it on his own, approaching not only Iran, but Algeria, Libya and other producing countries with the offer of secure outlets in Italy in exchange for concessions on their lands. Taking advantage of the Suez Canal crisis of 1956, he proposed to Iran, and successfully gained, an entirely new concessionary deal based on the revolutionary concept of a 50/50 partnership between ENI and NIOC in which all the expenses would fall on the former until oil was found, a temptingly riskless partnership for Iran.

'We aren't buried yet,' said an oilman as Mattei walked past him in a hotel lobby, 'but there goes the man who's driven the first nail in our coffin.'[5] From an oil company viewpoint, he was right. Mattei's path-breaking agreement was to open the door to a series of deals between Middle East countries and new entrants, including the Japanese and the American independents which allowed for state participation from the start of a discovery being made. The next company through the Iranian door was in fact Standard Oil of Indiana. But as long as these deals applied to new concessions and the major oil companies were left, albeit sometimes with tougher terms, with their existing ones, their stranglehold on supplies remained unaltered.

So Mattei discovered. Despite touting for new concessions all along North Africa from Morocco to the Sudan and despite eagerly drilling in his new Iranian concession, he didn't find the oil he wanted. 'An oilman without oil', as American executives called him, he turned instead to an even more politically explosive source – Russia. The USSR, which had not been exporting oil since the 1920s, was now keen to develop its hard currency exports to pay for the imports that its new leader, Nikita Khrushchev, wanted to help modernize the country. After a series of meetings between the two men, Mattei emerged with a contract to take Russian oil at low prices, which he then used to launch an aggressive price-cutting war aimed principally at Esso. Esso in turn was forced to reduce its official prices for Middle East oil to match the Russians, thus infuriating the exporting countries. Even then the American company found itself being consistently undercut à la Rockefeller. Taking advantage of the good offices of the American Government, headed by a president in John F. Kennedy who was eager to make a new start in European relations, Esso seduced Mattei into negotiations under which they were prepared to sell him cheap oil from their Libyan concessions in return for him giving up the war. Having being kept out of the majors' club so long, Mattei looked finally as if he were about to join them, when on 27 October 1962 his aircraft took off from Sicily in bad weather and crashed, killing all the occupants.

Mattei's death became as politicized as his life had been. A sabotage attempt had been made on him once before. The Italian press was quick to point the finger at the oil companies, forcing a rapid statement of condolence and expressions of goodwill from Esso. There were suggestions that the crash might have been engineered by the right-wing French faction in Algeria, with whom Mattei had fallen foul because of his dealings with the Algerian nationalist movement. There were alternative theories that the Mafia were to blame, concerned that Mattei's planned industrial investment in Italy might have reduced their hold on the local population. And then there was the possibility, of course, that it was just fate.

In a seminal article in *Politique Étrangère* in 1957, Mattei had argued that:

> . . . in concentrating in a few hands the control of oil production and marketing; in maintaining with the consumers the relations of supplier and customer in a closed and rigid market, in granting only financial returns to the countries that own the oil, and in barring all international agreements for rational organization of the market, the international companies have increased their own power, but they have also created the conditions for either the break-up or the transformation of the system under the pressure of new forces.

The break-up of the system was to take much longer than Mattei expected. For the drawback to his dream of a 'rational organization of the market' was that the interests of consumer and producer were not innately compatible. The consumers wanted cheap and secure energy, the producers wanted the highest prices obtainable and the maximum flexibility of manoeuvre. The Seven Sisters' sorority, 'the business in between producer and consumer', had its weaknesses and its banditry but it also had its uses, if only because the alternatives were so disruptive and so rarely successful.

Nationalism

The challenge came finally not from the consumers but from the producers. Relations between the oil companies and the producers were never happy, even at the beginning. No country with export commodities ever feels entirely secure with the organization that handles their exploitation, least of all of a commodity which provides over 90 per cent of foreign earnings and nearly as high a proportion of total government revenues as oil did in most of the Middle East and Latin American producing countries in the post-war period. When the concessions themselves had all too often been parcelled out as part of political deals between the imperial powers, their unease became even more understandable.

The nationalist revolutions in Mexico before the First World War had led to a complete state take-over of the industry in 1938, setting off a distrust between the oil companies, which boycotted Mexican oil, and the Mexican Government, which was to last for a generation after. Just after the Mexican move, and encouraged by it, a new military government in Venezuela used nationalization as a bargaining counter with which to force the companies to agree to much tougher terms, including a 50/50 profits' split. The concession terms in Iran, meanwhile, were revised after the Iranian Government had threatened to withdraw the concession altogether and the case had been sent to the League of Nations in 1933.

It was the war itself which helped spark off nationalism in the Gulf. Middle East oil became of vital importance and Germany was not slow to stir up Arab nationalism with promises of independence should it win. There were riots in Kuwait, where concessions had been granted but development had yet to get under way; a revolution took place in 1941 in Iraq during which the Iraqi Army for a time seized the oil installations in preparation for a German take-over which never came; and the British and Russians took over parts of Iran – the British the oilfields and installations of the south and the Russians the north – to ensure their interests against the pro-German Shah Reza, who was to be exiled at the end of the war.

Mossadeq and Iran

If Shah Reza was never that popular within Iran, nor were the Russians or the British. Nationalist sentiment found its idol in Dr Mohammed Mossadeq. Neither the Americans nor the British, nor for that matter most Iranians, have known quite how to take Mossadeq, then or since. Over 70 years old at the end of the war, a thin, frail figure often dressed in pyjamas, with an ailment which caused him to faint at the peak of emotion, often with calculated dramatic effect, alternately weeping, blustering and impishly playing with negotiators, this aristocratic rabble-rouser from one of Iran's most wealthy landed families seized on oil as the symbol of his patriotic fervour almost as soon as the war was over.

As a young lawyer and a member of the Majlis, or parliament, Mossadeq had played a part in the earliest days of revolutionary revolt against the Qajar royal rule in 1911–13 and in the efforts to create a parliamentary democracy in Persia under Shah Reza, the army officer who had taken power in 1921, declaring himself shah, or king, partly at the instigation of the British. In the twenties Mossadeq had been Governor of Fars, working reasonably well with the British presence there and with Anglo-Persian until, refusing to vote in favour of Shah Reza's demands for dictational powers, he was arrested in the thirties.

Shah Reza proved too much for the British also. Exiling him in 1941, he died under their restraint in South Africa three years later. Released, Mossadeq became a national figure, prominent in the movement at the end of the war to rid the country of both British and Russian troops and to overturn the oil agreement which the Russians had attempted to impose on the northern part of the country. The new shah, Shah Reza's son, was young, inexperienced and vacillating, driven by a strong-minded sister, Princess Ashraf and a genuine desire to bring some pride back to the country, but conscious of how much he owed his position to the British and the Americans. For a time the Majlis emerged as a relatively free and voluble force in which Mossadeq became a prominent leader. In 1944 he helped carry through a bill, aimed largely against the Russians, making it illegal for a government minister or official to negotiate oil concessions without the permission of the Majlis. In 1947 the Majlis rejected outright a concession forced out of the Iranian Government by the Russians, declared its intention of forming new petroleum laws and, in a clear warning shot against Anglo-Iranian, announced that:

> . . . in all cases where the rights of the Iranian nation, in respect of the country's natural resources, whether underground or otherwise, have been impaired, particularly in regard to southern oil, the Government is required to enter into such negotiations, and take such measures as are necessary to regain the national rights, and inform the Majlis of the result.[6]

Anglo-Iranian, now headed by a dour and dedicated Scot, Sir William Fraser (later Lord Strathalmond) who had come to the company when it had taken over his family shale business in Glasgow, took this merely as a piece of rhetoric that required a pounds, shillings

Above: 'Colonel' Drake (*left*), whose discovery of oil in 1859 began the oil era. James Young (*right*) founded an industry by refining oil from shale but his methods were superseded by Drake's extractive techniques.

Below: A forest of derricks at Power Run Creek, Pennsylvania. The oil rush sent prices tumbling from $20 per barrel to 10 cents within a few years.

Above: G.B. Reynolds sharing lunch with two colleagues during his three-year search for oil from 1905 to 1908 in the inhospitable mountains of southern Persia.

Below: The Shah of Persia (*left*) who awarded the concession to William Knox D'Arcy (*right*). D'Arcy spent nearly all his fortune financing the Persian search. Although he founded Anglo-Persian, he never visited the country before oil was discovered there.

Above: The signatories to the Achnacarry Agreement of 1928. Sir Henri Deterding (*left*), challenged John D. Rockefeller for world dominance of the oil industry but preferred co-operation to competition as the means to order. Meeting with him at Achnacarry Castle in Scotland were Lord Cadman (*top right*), chairman of Anglo-Persian, and Walter Teagle (*bottom right*) of Standard Oil (New Jersey), the main inheritor of Rockefeller's Standard Oil empire.

Left: Sir Marcus Samuel, later Viscount Bearsted. More interested in the pomp of London than the circumstance of oil, Samuel lost out to Deterding in the Royal Dutch/Shell merger of 1906.

Above: Challenge to the power of the oil companies came from Mohammed Reza (*top left*), Shah of Iran from 1941 until 1979; Dr Mossadeq (*bottom left*), Prime Minister from 1951 to 1953; and in Italy from Enrico Mattei (*right*), head of the state oil company, ENI.

Below: Senior executives of the US oil majors swear the oath before Senator 'Scoop' Jackson's Permanent Subcommittee on Investigations during the oil crisis of 1973–4.

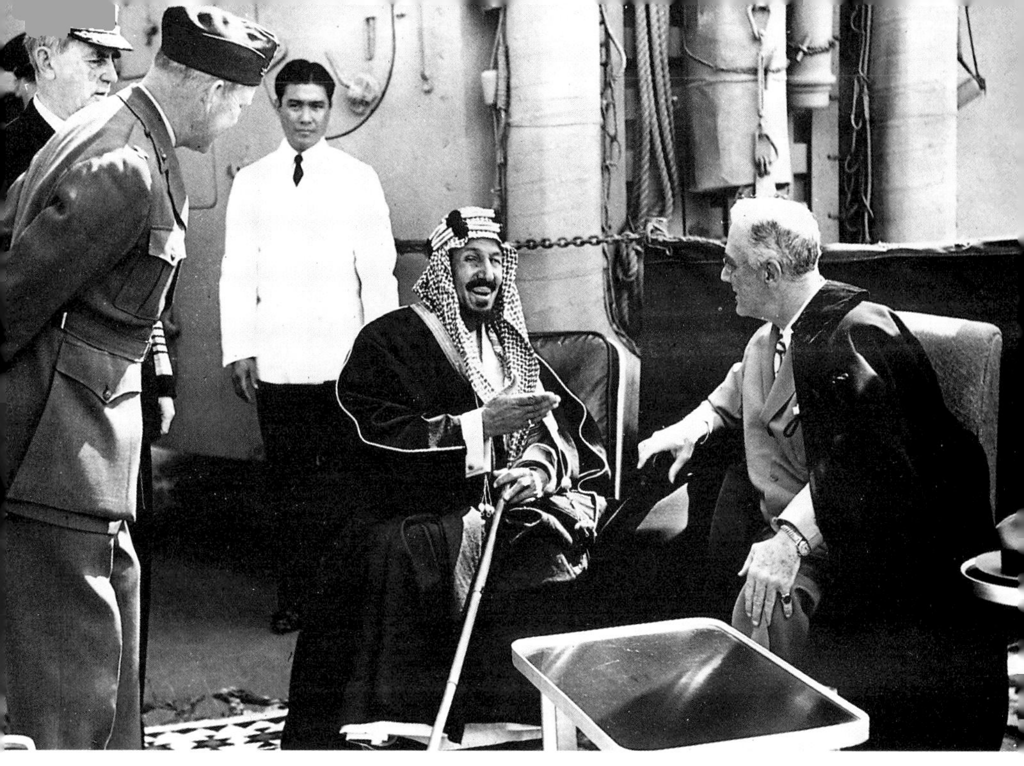

Above: King Ibn Saud meets President Roosevelt, aboard the cruiser *Quincy* in Egypt in 1945, setting the seal of American post-war influence in the Middle East.

Below: Sheikh Yamani at an OAPEC meeting in October 1973. The decision by Arab oil ministers to embargo oil shipments after the Arab-Israeli war set off the first oil crisis.

Above: 'Dad' Joiner (1) congratulates geologist, Dr A.D. Lloyd (2) on the East Texas oilfield discovery, while into the picture steals H.L. Hunt (3).
Right: Hunt's sons, Herbert and Bunker, are sworn in before a House Committee in 1980 investigating their efforts to buy out the world's silver.

Below: Paul Getty waits to hear his bids read out at the auction of UK offshore leases in 1971.

Above: T. Boone Pickens, head of Mesa Petroleum, today's oil millionaire who has made his fortune by 'drilling for oil on Wall Street'.

Below: The New York Mercantile Exchange which made trading in oil a market for the speculator and the investor and revolutionized the way oil prices were fixed.

A semi-submersible rig dwarfs Lerwick harbour in the Shetlands, a symbol of the power of oil in a small community, promising riches and threatening to overwhelm islands once dependent on sheep, fishing and emigration for their livelihood.

and pence settlement. It had happened before when Shah Reza had withdrawn the concession in 1933. Then everything had been sorted out behind closed doors and in palace corridors. The same could be done again. Used to a life of some colonial splendour in south Iran, an area which had been virtually independent of Tehran control for most of the century, Anglo-Iranian officers and British embassy officials tended to regard the Majlis as weak and corrupt and the Iranians as basically inefficient and mercurial. They also took the view, as did company after company in the decades of nationalism and oil renegotiations that were to follow, that the Middle East was a single entity. You had to avoid giving in to pressure at any point in the region, however slight, lest it prove the thin edge of the wedge for concessions to be granted in another.

Nevertheless, although Fraser has been much criticized since, not least by American oilmen and the younger generation of Arabists who were just climbing up the British Foreign and Colonial Office hierarchy, the company was not entirely blind to the need to take account of the new world. The more rhetorical concerns of the Majlis were suddenly augmented by a threatened financial crisis when the British Labour Cabinet severely reduced Iranian oil revenues by limiting the dividend payments of British companies. The Government's action was entirely accidental in its effects on Iranian oil. It was intended to control profits and prices at home. As 20 per cent of Anglo-Iranian's dividends over a set sum were due to the Iranian treasury, however, its effect was sharp. During 1948–9 negotiations took place to arrange new financial arrangements which would guarantee Iran a higher income from royalties and profit-sharing. The Supplemental Agreement as finally signed was by the company's own reckoning, a generous one. Unluckily for Anglo-Iranian, the agreement was finally presented to the Iranian Majlis in January 1951, only a month after the Aramco partnership had agreed a much more ostensibly attractive 50/50 profit-sharing agreement with the Saudi Arabian Government, although in fact this offered less in real terms. It was Anglo-Iranian's – and Britain's – added misfortune that the agreement was put forward at a time when Iran's internal political situation was thrown into confusion by the assassination of the Iranian Prime Minister, General Razmara, by a Muslim fanatic in March 1951. Razmara had been appointed, with British approval, as the strong man who had to bring order to the situation. His death left an indecisive young shah with little alternative but to turn to Mossadeq. Although head of only a small faction in the Majlis, Mossadeq had the backing of the landowners, who wanted to reassert the authority of Parliament against the Crown; the Communist Tudeh party; and the students and the religious mullahs, who resented the West and its introduction of foreign ways.

With great rejoicing the crowd and the Majlis acclaimed Mossadeq. His first act was to declare his intention of carrying out the nationalization of the oil concession. 'He was an old-fashioned farmer,' recalls Sir George Middleton, then just arrived as an adviser at the British Embassy in Tehran, 'who'd come to market and seen that apples were selling at 25 cents a pound and he was only getting $2 a bushel. And he went mad. . . .He didn't understand about the market, but he just felt that he was being robbed and, as an old-fashioned farmer, he wanted to put it right.' He was also a shrewd politician. For oil more than any other issue caught the ever-present and fervent nationalism of the Iranian people and aroused their profound sense of humilitation at its control by what they saw as an agent of the British Government.

Losing no time, as soon as he was officially appointed Prime Minister in May 1951, Mossadeq sent down the Governor of Khuzestan, accompanied by a crowd of enthusiastic supporters, to the Anglo-Iranian headquarters at Kermanshah to take over. The trouble was that, if Mossadeq saw the situation in simple terms of retrieving what was rightfully the country's and should never have been given away, the company saw it equally clearly as a case of illegal seizure of an asset for which they had honestly negotiated and for which they would have to be properly compensated. Add to a feeling of absolute legal right, a sense of colonial superiority among the staff and their wives, and there were all the makings of an impossible confrontation.

Mossadeq appeared to think that it was just a matter of taking over the installations and the Abadan refinery (then the largest in the world), continuing to pump out the oil and then negotiating such compensation with Anglo-Iranian as was necessary. The British employees, he naively assumed, would just follow the orders of their new masters, content to keep their luxurious lifestyle. Not surprisingly, the British managers of Abadan and their chiefs back in London didn't quite see it that way. Led by Eric Drake, then general manager on the spot and later to become chairman of the company, they refused to co-operate, keeping back the files, avoiding crucial maintenance and losing vital parts to machinery. More effectively, the company back in London got in touch with all potential buyers of the oil to warn them off. Mossadeq was insisting that any tanker captain loading oil must sign a receipt stating that the oil was Iran's. The company said that it would take immediate legal action if they did. 'Whoever bought Iranian oil brought a lawsuit with it,' as Sir Hartley Shawcross, then Attorney-General, put it.

Only three tankers loaded oil – two Italian, one Japanese – and each arrived home with a writ upon it. The oil companies, who had been through this before with the Russians and the Mexicans, were all too conscious of the need to keep together under such circumstances, to challenge Anglo-Iranian at least in so far as legal actions were concerned. The tanker owners, for their part, were far too dependent on the oil companies to risk their wrath by picking up speculative oil. Within a month all exports from Iran had ceased and with the storage tanks full up, the wells and pipelines had to be shut down. The company wives were sent home and Drake, feeling himself threatened by a law promising the death penalty to anyone refusing to co-operate, skipped over the border to Iraq and then moved offshore to a ship in the Gulf from which he could watch how the stalemate developed.

The answer was, not at all smoothly. Nationalization of oil was a political issue which posed severe problems. Mossadeq was not prepared to compromise. The question of sovereign rights wasn't, he felt, the kind of question he could compromise over. Nor, given the popularity of the action as a symbol of Iranian resurgence, could he look as if he was nickel-and-diming a solution. Meanwhile, the Shah vacillated, fearful of the public reaction if he tried to unseat Mossadeq.

The British Labour Government was in a quandary. It had nationalized far too much itself to object to the principle of a nation taking over its main asset. It was also anxious to appear the friend of aspiring nationalists. Oil was, however, vital to the British economy. Not only did Anglo-Iranian represent the country's biggest single foreign investment, but its control of Iranian production also ensured that oil was bought in sterling to the immense benefit of the much-troubled balance of payments. George Middleton's instructions on taking up his post in Tehran were simple: 'Keep the oil flowing and make sure it's in sterling.'

The Foreign Secretary, Herbert Morrison, and the Defence Secretary, Emmanuel Shinwell, wanted action. So did the newspapers who, encouraged by searing criticism of their inaction by the leader of the Opposition, Winston Churchill, suddenly became enthused with one of their periodic bouts of jingoism. Eric Drake flew back to London on his own initiative amid an atmosphere of great secrecy, false names and terror of the press.

He arrived to find, on his own account, a cool lack of interest from his chairman, Lord Strathalmond, who appears to have been both irritated by Drake's original exit from his post and even more furious at his unauthorized arrival in Britain. This was not the case with the Government. Within days of his arrival, Drake was summoned to a Cabinet meeting to address both ministers and the heads of the armed forces. He pleaded with the Government for action if the principle of international commercial rights was to be protected and if British interests were to be preserved. If the British let this one pass, Drake warned, the the Suez Canal would be nationalized within five years. He was right, as he now proudly claims, to the year.

The difficulty for the Cabinet was that, as their successors were to find in the Suez crisis, the old days of gunboat diplomacy were passing.

It was no longer possible to consider military action without American approval. And the Americans, advised by an ambassador sympathetic to Mossadeq and headed by an administration under President Harry S. Truman with a good deal of natural sympathy for the nationalist aspirations of new countries in revolt against the British Empire, did not approve of sending in the troops or the Navy.

The British Government huffed and puffed. A destroyer was sent to evacuate British personnel and position itself ready to bombard the Iranian coast. The troops in Cyprus and the airbase in Iraq were reinforced. Robin Zaehner, an academic and former intelligence officer in Iran during the war, was sent out with instructions to scout out the political landscape and to root out potential opponents to Mossadeq, a task he performed with the help of three *bazaari* merchant brothers, the Rashidians. In the end, however, Prime Minister Clement Attlee refused to press the trigger for military action. Influenced partly by American advice, he sought a solution through negotiation and through the United Nations the courts. The courts were unhelpful. The International Court in the Hague, after months of consideration, finally decided that it had no jurisdiction. Mossadeq flew to the United Nations, where he readily persuaded most of the Third World to support his David-like struggle against the British Goliath. Negotiators sent out by both Britain and the United States – the experienced Averell Harriman on behalf of Truman and Richard Stokes on behalf of Attlee – and both ended their trips to Tehran frustrated but largely sympathetic to the Iranian case.

British attitudes were toughened against Iran by a change in government in October 1951 which brought in Winston Churchill, who had been belabouring the Government for running away from the crisis, as Prime Minister, and Anthony Eden, himself an orientalist and fond of quoting Persian poetry in his more fanciful moments, as Foreign Secretary. Eden was not unsympathetic to Iranian historical aspirations but equally he felt on firm enough ground himself to take a more muscular view of the possibilities of replacing Mossadeq, if only the Shah (whom he had helped to put on the throne) would co-operate.

C. M. 'Monty' Woodhouse, an enthusiastic intelligence officer for MI6, the British equivalent of the American CIA, was sent out to build on the network started by Robin Zaehner and the British Government started to lean heavily on the Shah to act. The favoured politician was Qavam Salteneh, an elder statesman sympathetic to British interests. The appropriate moment came in July 1952 when Mossadeq, who was becoming increasingly dictatorial as international and economic pressures mounted, demanded the right to appoint his own War Minister. The Shah refused. Mossadeq resigned and Qavam was appointed in his place.

The peaceful coup was a disaster. In a foretaste of what was to come twenty-seven years later with the deposition of the Shah, the Tehran mob rose up in demonstrations against the new Prime Minister. Many of the troops that were brought in refused to fire on their compatriots. Qavam resigned and the Shah was left exposed to Mossadeq's revenge. His sister, Princess Ashraf, was sent into exile. Mossadeq made himself War Minister and obtained from the Majlis the powers of a virtual dictator.

The Demise of Mossadeq

From then on the political situation deteriorated rapidly. The more isolated Mossadeq became internationally, the more authoritarian and unyielding he became at home, ignoring the Majlis, leaning increasingly on the crowd and the Tudeh Communist party for support and proving endlessly resistant to American efforts to intermediate. 'It was like walking in a maze,' recalled Dean Acheson, Truman's Secretary of State, 'and every so often finding oneself at the beginning again. On price particularly he pretended to be very vague and stupid.'[7] When the World Bank offered its services as mediator in the dispute, Mossadeq replied that he would only accept if they announced that they would be representing Iran, thus making negotiation with the British impossible. In October 1952, aware of Monty Woodhouse's efforts to foment rebellion and opposition, Mossadeq broke off diplomatic relations with Britain altogether, accusing the British Embassy of spying. The following month the wider political environment altered very much in his disfavour, however, when General Dwight Eisenhower was elected President of the United States.

Eisenhower was pro-British, deeply concerned at the break-off of diplomatic relations and, in tune with American popular opinion at the time, sharply anti-Communist. When Monty Woodhouse, ejected from Iran and recommending that the time was right for a coup, reported back to Anthony Eden, he was instructed to try to gain Washington's support for the idea. He flew there to see his opposite number at the CIA. On his own account, the plan he took was very much 'a Hollywood affair' of a dramatic military action to take over Tehran, arrest Mossadeq and instal a more sympathetic government. However, by playing the threat of Communism card for all it was worth, he interested the Americans in the plan. 'One thing that became fairly obvious to me right from the beginning,' Woodhouse recalls, 'was that there was a sharp cleavage in point of view between the American oil industry and the American Government. . . .The important thing seemed to me to be to convince the Americans that what was at issue was the danger of Iran going through a revolutionary process somewhat similar to what happened in Afghanistan a quarter of a century later. In other words, first of all a nationalist revolution then a take-over of the nationalist revolution by the indigenous Communists and finally a take-over of the Communist revolution by the Soviet Union.'

Oil company advice to the American Government was unemphatic. The American oil companies, with their own interests at stake, argued that Iran was not an issue for the United States to get too involved in. The world could manage without its oil. It was a British not an American affair. Where they were emphatic was on the Communist question which was critical. Their advice fell on ready ears. Kermit Roosevelt, grandson of former President Theodore Roosevelt and then head of the CIA's Middle East operations, went into action in a classic plot of internal subversion. A new prime ministerial candidate, in the form of the ambitious General Zahedi, was picked, slightly against British sensibilities as they had marked him during the war years as a pro-German sympathizer. A network, partly using the Rashidians, was established and now the Americans started to lean on the Shah.

Having been bitten once, the Shah was in no mood to lead another failed coup. It took a good deal of time and various messages from the British and American governments including the use of code words in an Eisenhower speech and a BBC broadcast, to convince him that this time both the Americans and the British were fully behind the removal of Mossadeq. Mossadeq himself in the meantime seemed intent on maximizing the opposition and minimizing the chances of second thoughts among the Allies. He announced his intention of abolishing the Majlis. His contempt of the Shah became greater, his absolute refusal to negotiate an oil settlement more dogmatic and his relationships with the Tudeh party more open. In July 1953 Kermit Roosevelt moved to Tehran. The Shah finally agreed to sign the orders to dismiss Mossadeq and to appoint General Zahedi in his place. He then dashed straight for his palace on the Caspian, where the orders had to be brought for his signature, and ordered an aircraft to be on stand-by, ready to speed him out of the country should anything go wrong.

It did. When the messenger bearing the orders arrived at Mossadeq's office, he had him promptly arrested. The crowd came out in Mossadeq's support and some of the army barracks in the capital rose up in his favour. The Shah fled to Rome. General Zahedi took refuge in the American compound. Roosevelt established himself in a temporary radio station, orchestrating counter-demonstrations of crowds, paid and genuine, to take to the streets in support of the Shah. The renting of mobs, and the payment of *agents provocateurs* by the CIA to shout Tudeh slogans while stoning mosques may have influenced events.

'I started off in Tehran, staying up in the mountains above the city,' remembers Kermit Roosevelt, who remains contemptuous of the British efforts before his arrival, 'and I did not move down to the town until the day of the operation was about to come. Then I moved into a little shack in the embassy grounds, where we, incidentally, reproduced copies of the orders dismissing Mossadeq and appointing Zahedi in his place and we gave them to our principal agents to distribute around the city, which they did very successfully. They were able to organize a rather impressive movement from the bazaar quarter of Tehran towards Mossadeq's house. It was led by the great weightlifters, the *zirkhani*, who couldn't take their weights with them. But they took batons and twirled them around and behaved as if they were jugglers instead of weightlifters. And this made a rather impressive procession marching on Mossadeq's house, which was calculated to swing support from the Army and the general population for the Shah and against Mossadeq, which it did.'

The most significant factor, however, was simply that Mossadeq was losing support. The Army, fearful of what would become of it in a popular revolt, swung the way of the Shah and the military commander of Kermanshah advanced on the capital with tanks. The *bazaari* (the crowds from the bazaar), turned on Mossadeq, inflamed partly by the mullahs, fervent in their fear of Communism, and the Tudeh party. The landowners and middle class of the Majlis had lost faith in Mossadeq because of his refusal to negotiate on oil and his failure to sort out the economic mess into which the country was sliding.

After days of violence and havering fortunes, the mob marched on Mossadeq, surrounding his house and rushing an armed guard in an affray that cost the lives of 200 and caused injuries to some 500 more. Eventually Mossadeq was arrested in his office in the Majlis, to be put on trial and sentenced to three years' imprisonment. It was a show trial which he played to the maximum. General Zahedi, for his part, emerged from hiding to take command of the city and send a message to the Shah to return, if not quite in triumph, at least to power.

Post-Mossadeq

Iran could not, however, be the same, nor could the oil industry, as it had been before Mossadeq. It was the Americans who had made the country safe once more 'for democracy' and it was the Americans who would have to take the lead, and a major share, in any oil settlement. As the American and British governments discussed the issue, they sought a political solution that would tie in the major Western interests with the oil production of Iran. A consortium of companies, including American, Dutch and French but not, to Mattei's fury, Italian, appeared as the logical route.

Anglo-Iranian, who argued bitterly for the maximum recompense to itself, was inclined to blame the solution to the crisis on the devious machinations of the American majors. This was only partially true. The American companies at the time had been enjoying the fruits of greatly expanded output from Saudi Arabia, Iraq and other parts of the

Gulf to make up for lost Iranian exports and were none too keen to see Iranian shipments coming back too strongly to the market. But on the other hand they were also facing energetic investigation and anti-trust action from the Federal Trade Commission and were wary of involving themselves in group action that would only add to the evidence against them.

It took some tough dealing by the State Department and the White House to rope back the Federal Trade Commission and some equally tough dealing by the British Government to keep Anglo-Iranian in line. Eventually it was decided that a consortium of the Seven Sisters plus CFP of France would be formed. A negotiating team was set up under the chairmanship of Howard Page, a veteran director of Esso who had just completed a series of successful and particularly difficult negotiations between the Iraqi Government and the Iraq Petroleum Company.

The talks with Iran did not proceed quite as smoothly as the British and American governments at least had expected. Conscious of the humiliations of his flight abroad and his American-controlled return, the Shah was now determined to assert Iran's national interests and not simply to be a pawn of the Western governments. He chose as his Oil Minister and chief negotiator one of the most urbane and skilled of diplomats, Dr Ali Amini, whom Howard Page remembers as 'one of the smartest people, if not the smartest, I have ever met in my life'.

The Shah called in General Zahedi and Dr Amini and, as Amini recalled, told them: 'The Americans are coming in search of an easy bargain. Anyone who gives it to them will answer to me and to the Iranian people. I want it known as widely as possible by my people that the oil will stay Iranian, and that they will decide the punishment of any negotiator who fails to bear that in mind.'[8] He meant what he said. Just as the first delegation of company executives before the committee under Page was set up, was leaving the airport to report back to its headquarters, it was handed a memorandum from the Iranian Government demanding full sovereignty and absolute national control.

It took four months of hard and continuous bargaining before agreement was reached in October 1954 with the help of the respective representatives of the American and British governments, Herbert Hoover, jun., son of ex-President Hoover and Eisenhower's oil adviser, and Denis Wright, the new British Ambassador to Tehran. The final compromise, discussed beside running water to avoid bugging, gave Iran full title to the oil and the refinery at Abadan and cast the eight-company consortium in the role of off-takers, guaranteeing to take a minimum quantity of oil each. The consortium was divided as follows: 40 per cent to Anglo-Iranian, which now renamed itself British Petroleum or BP to mark the historic change in its circumstances; 14 per cent to Shell, 6 per cent to CFP and 8 per cent each to Esso, Mobil,

Socal, Texaco and Gulf. The American companies were forced within a year to give up 1 per cent each to buy off a group of nine independent American oil companies, which were complaining bitterly of being left out and who were threatening to go to the Federal Trade Commission unless they got a share of the action. BP managed to gain over £500 million in compensation to be paid directly by the Iranian Government and indirectly by the other members of the consortium as their fee for gaining a part of its original concession. The Shah obtained a restoration of oil shipments at a minimum level. Only Mattei was left out in the cold, in a slight which the oil companies were later to regret.

The agreement with the National Iranian Oil Consortium (NIOC) was to last, like the Shah, for almost twenty-five years, in formal terms at least. Although it gave the British a considerable shock and raised sudden fears of a wave of national revolution throughout the Middle East (the deal occurred at the same time as Nasser came to power in Egypt), it showed emphatically that the oil industry and the consumers could, if need be, manage without a major source of Middle Eastern oil – a point that was to be re-emphasized during the closure of the Suez Canal in 1956 and again in the Arab-Israeli War of 1967. Oil companies were able simply to increase the supply from other countries to make good the difference. The price of their help was a consortium and an end to one-company dominance of Iran. The price of the Shah's acceptance of the reinstatement of the oil companies was guaranteed off-take, a point that was to cause constant friction with the Iranian Government over the ensuing decade. As far as the rule of the Seven Sisters and the Anglo-American alliance, formed to ensure security of supply, was concerned, it confirmed the pattern for the next decade. In all the talks between the British and American governments following the fall of Mossadeq, what came through most clearly was the primary concern of the European consumers in the security of oil supply over and above its price. And that security seemed, at the time at least, to be best ensured by the strength and diversity of the international oil majors.

When details of the secret arrangement, the 'Aggregate Programmed Quantity', under which the consortium members parcelled out their needs of Iranian oil each year on the basis of the minimum requirement of each company, were revealed nearly twenty years later by yet another US Senate investigation of the oil industry cartel, the details infuriated both the investigators and the Shah. The idea that oil companies should secretly fix between themselves as a matter of commercial discretion a matter so important as the annual output of Iranian oil seemed deeply offensive to the Iranian Government and to the American public. It was not how the companies saw it themselves, though, then or since. As Howard Page, for example, views it even today, the arrangement was simply a means of sorting out otherwise incompatible demands by the Middle East producers, each of whom wanted the majors to take more of their oil than their neighbours. To the Seven Sisters, it was managing the business in a balanced way. The consumer governments appear to have agreed. Better order than chaos and far better the order of companies than their own direct involvement in oil relations with the producers, was their instinct.

The Fall of the Shah

For the next twenty years the Shah of Iran was to pressurize the oil consortium constantly either to take more oil or to pay more for it. In a number of ways he proved far tougher in negotiation and far more ruthless in pursuing Iranian interests on oil matters than Mossadeq had been. His brief exile, his third marriage to Queen Farah Diba and the birth of a son and heir, coupled with a shrewd and wily brain in handling court politics brought out in him qualities of leadership that surprised those who had known him in the early days. He always, however, conducted his oil negotiations in pursuit of commercial gain for his country rather than seeking the means of controlling it as a national resource. The lesson he appeared to have learned from Mossadeq's failure was that oil was too important to be left to idealism and that it should be used to buy jobs, progress and industrial pride rather than as a focus for nationalistic slogans. For a long time that judgment seemed to be correct. Iran's policy led the way not to national revolutions but to obtaining greater wealth from the existing company structure. And yet, if the Mossadeq revolution came to seem a peculiarity of the very separate history of Iran, to those who looked closely enough, it none the less gave due warning of the far greater cataclysm that was to sweep the country and the whole Islamic world twenty-five years later when the students and the *bazaari* took to the streets in the name of an even more elderly figure, that of the Ayatollah Khomeini.

Khomeini, like other religious leaders, had kept his distance from Mossadeq during the latter's final year in office. The mullahs of Iran's predominant Shiite Muslim sect, had become too distrustful of Mossadeq's intimacy with the Tudeh party. They were no happier, however, with the Shah and his view of oil as the vehicle with which to modernize Iran along Western lines. In 1962 Khomeini emerged as the leader of the religious opposition to the Shah's plans for land reform and the emancipation of women. In the following year he was exiled to Iraq, whence he finally moved to France.

From the left-wing, or liberal, standpoint – the platform for most of the articulate opposition to the Shah – the Ayatollah should have been an anachronism. Indeed there were many, even among the religious leaders of the time, who were quite happy to see him abroad. But he caught the current that was to grow in strength and speed as oil brought new wealth to the Middle East: the distrust of its Westernizing influence. The Shah had hoped that by spreading oil wealth more widely and creating a new middle class of technocrats and industrialists dependent on economic progress for their own livelihood, he would cement a major part of the population behind royal rule. That was the analysis of a number of foreign experts too, such as Sir Anthony Parsons, British Ambassador to Iran in the final years of the Shah's rule. Previous regimes, he was fond of saying, had been brought down in the Middle East not by the force of opposition but because of the lack of support. The Shah was different. Even if he was felled by the assassin's bullet, his regime would live on. There were too many interests dependent on it.

Under ordinary circumstances that might have been so. Instead, the opposite happened. The new educated elite, and their Western-educated children, including the daughters who might have been thought to benefit most from the Shah's policies, reacted strongly against his growing authoritarianism and the corruption of his court. Khomeini had been arrested in the holy city of Qom in 1963 only with a massive use of force and around 1000 dead. The secret police, the *Savak*, became increasingly powerful as the Shah became more afraid of the bullet and the number of clandestine parties established to overthrow his regime by force. What should have been the climaxes of his rule: his enthronement in 1967, the celebration of 2500 years of the Persian monarchy at Persepolis in 1971 and the fiftieth anniversary of the Pahlavi dynasty in 1976, proved to be irritants, reminders to the opposition of the Shah's excessive spending and the fawning of foreign powers. His abolition of all political parties save his own, his effort to reform the calendar around the coronation of Cyrus the Great and his appropriation of the special title of *Aryamehr*, 'Light of the Aryans', were meant to demonstrate the superiority and glories of Iran. Within the country they were all too often seen as the empty gestures of a man suffering delusions of grandeur.

Worse, the oil boom on which the Shah had founded his 'White Revolution' and his ambition to make Iran into a great industrial nation, turned sour. The Iranian economy overheated, particularly after the quadrupling of oil prices in 1973. Inflation soared. The huge investment plans in new plant and industrial complexes undertaken by the Shah quickly faltered on the bottlenecks of poor port facilities, inadequately trained technical staff, lack of management and corruption at the top. The Shah was forced to revise his plans and to intervene on prices, thus upsetting the commercial interests and the merchants whom he had looked on originally as his main allies in his fight against feudalism. As the towns became filled with immigrant tribesmen without jobs a new category of urban poor was created.

The Americans, who had replaced the British as the main supporters of the regime, tried to soften the temper of the Shah's rule. Like the British before them, however, they were caught by their own international policy imperatives. The Shah was seen as the mainstay of Middle Eastern defence against Communism. His purchase of arms and capital goods from the United States had become an essential means for the West to pay for the higher price of oil. Henry Kissinger, President Nixon's National Security Adviser and later, Secretary of State, was almost embarrassing in his praise. President Jimmy Carter, who had come to office preaching the pursuit of a more moral foreign policy, greeted the Shah on the tarmac of Washington airport as America's closest ally.

When the end did come at the beginning of 1979, the Americans had neither the forewarning nor the will to intervene. In common with the British and other European embassies, they had grossly underestimated Khomeini's strength, just as they had once underestimated Mossadeq's. And just as the moderates they promoted to stem the tide, like Qavam Salteneh, had failed then, so the moderates such as Dr Bakhtiar, were to fail now. Once again the power of the populace proved irresistible. This time the Shah was forced permanently into exile by the mob. The era had passed when the CIA could form the counter-revolution.

The event which had set off the Shah's downfall was a strike by the oil workers in November 1978 which deprived him of his financial lifeline. The new leaders who came in not only challenged the oil industry and its Western domination. They did worse. They suggested that oil had done the country nothing but harm.

4

'Charter for Change'

'From the early 1960s until the early 1970s,' recalls Sheikh Ahmed Zaki Yamani, Saudi Arabia's elegant Oil Minister for twenty-four years, 'the price of oil was determined by the oil companies. In the early seventies the price of oil was determined by both OPEC and the oil companies. Then from 1973 until 1982 or 1983, OPEC played the role of pricing oil on its own. They overdid it between 1979 and 1981. And that was a mistake and we're paying for that mistake. We became weak again.'

More than most, Sheikh Yamani has symbolized the dramatic swing in the balance of power between companies, producers and consumers that has made oil so explosive an issue in the modern age. Only a few years after the founding of the Organization of Petroleum Exporting Countries in 1960, Sheikh Yamani became and has remained its voice and its most acceptable face, explaining the organization's role and his kingdom's views at countless press conferences and interviews. Whether announcing a doubling of oil prices in 1973, depreciating the extent of a further doubling in 1979, urging the oil companies to show sense in accepting Saudi demands for state participation in 1972 or announcing his kingdom's refusal to accept the role of swing producer for OPEC any longer in 1985, Yamani has always been the epitome of patient reason.

The story behind these events has been partly OPEC's as the producers have tried, with varying degrees of success to ride the tiger of the oil market in the last quarter of a century. More particularly it has been that of the Persian Gulf and Saudi Arabia. As an American geological mission, led by one of the great pioneers of geophysics and oil exploration, Everett Lee deGoyler just after the Second World War wisely concluded: 'The centre of gravity of world oil is shifting from the Gulf–Caribbean areas to the Middle East, to the Persian Gulf area, and is likely to shift until it is firmly established in that area.'[1]

The Middle East had, and still has, the reserves. Even today, when high prices in the seventies and early eighties have induced oil companies and governments alike to search for every smidgen of oil from New Mexico to New Guinea, nearly 60 per cent of the world's proven reserves lie in the Middle East and nearly half of these are in Saudi Arabia. Costing 10–30 cents per barrel to produce, Middle Eastern

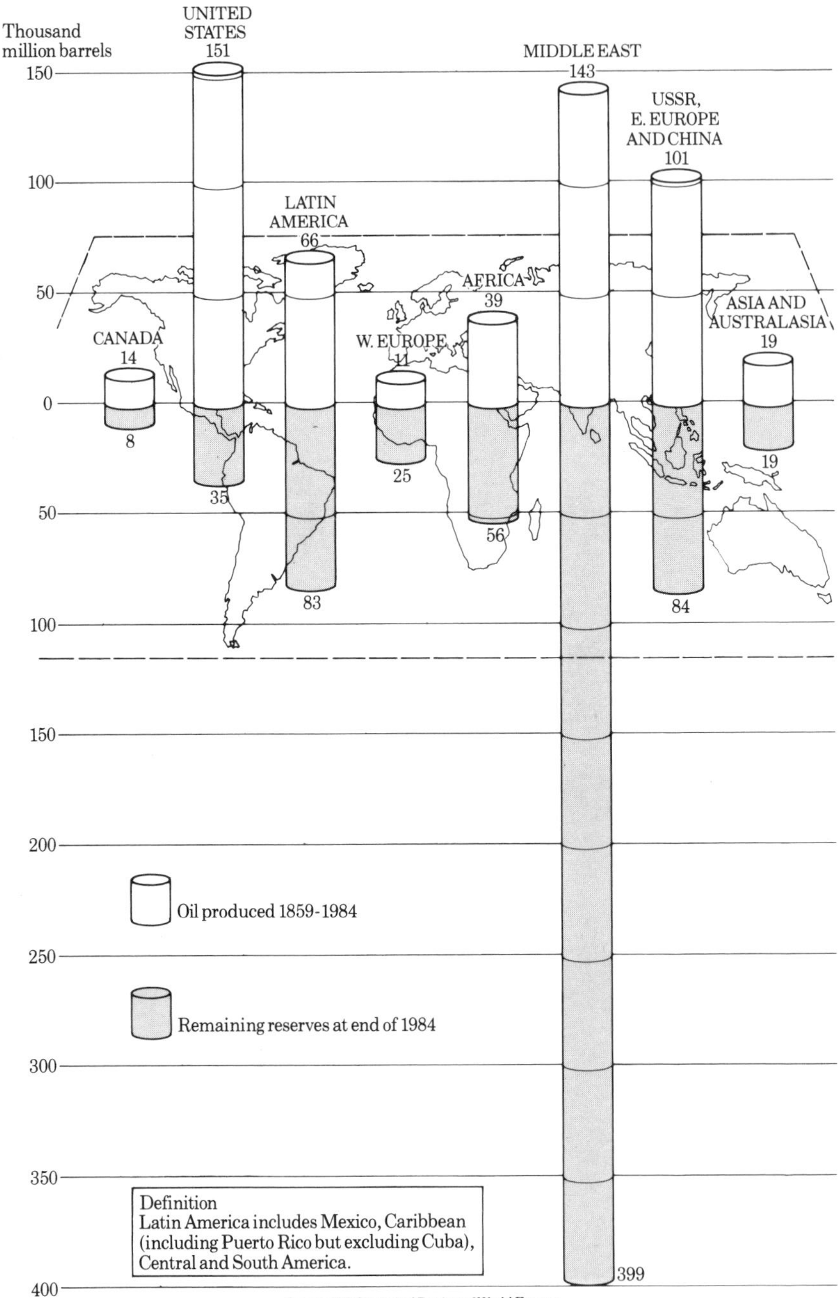

Cumulative world production and remaining reserves in 1984. The chart shows how far the United States has already used up its reserves, while the Middle East still holds more reserves than all the rest of the world put together.

oil is by far and away the cheapest and there is enough at present rates of production to last over 100 years, ten times longer than American reserves and five times longer than those of the North Sea.

When oil was first discovered in the Middle East by Reynolds and the other early drillers, it was in territories at the far reaches of the old Ottoman Empire, governed by local emirs and sheikhs, populated by nomadic tribesmen and local fishermen and held under British influence by the threatening presence of her fleet and her political advisers in most of the local courts. The balance of oil power organized at the time lasted well into the post-war period. Persia was under the control of the British and the Anglo-Persian Oil Company (later called Anglo-Iranian to fit in with Shah Reza's retitling of the country in 1935). Iraq had been conceded to the Turkish Petroleum Company, or the Iraq Petroleum Company as it had been renamed in 1929, its hold actually increased by the granting of all the country's oil rights in 1938. As the world emerged from war it was these two countries that led Middle East production with around 80 per cent of a then relatively small output of 26·5 million tonnes in 1945.

Production and pricing, however, were still bound by the constraints laid down by the major oil companies in the Red Line and Achnacarry agreements of 1928. Of the seven original American companies which had joined the Iraq Petroleum Company, only Standard Oil of New Jersey and Standard Oil of New York remained partners by the end of the Second World War. Texas and Sinclair dropped out before negotiations were completed. Atlantic and Pan American (later a subsidiary of Standard Oil of Indiana) sold out early on, only to buy their way back into the Middle East in the 1960s at far greater cost. Gulf eventually resigned to pursue an independent policy, free from the agreement's restriction on any member exploring for oil by itself within the Red Line area.

Playing the Arabs at their own Game

The country Gulf picked was Kuwait, a state excluded from the agreement. The company was led there by way of Bahrain by a New Zealander, Major Frank Holmes, one of the more colourful middlemen of the early years of Middle East oil. A stocky man with rough manners and an open disposition, Holmes was a self-educated mining engineer who, having served with the British forces in Mesopotamia in the First World War, stayed on in the area to work his way as a concession promoter.

Sporting gear eccentric even by the standards of Middle Eastern explorers, Holmes, with his white helmet covered in a green gauze veil and carrying a white umbrella also lined in green, turned up at the court of Ibn Saud of Saudi Arabia in 1923 looking for mineral rights. Much to the irritation of the British, who wished to keep the Arabian peninsula under their own political direction and commercial development, with the help of a Lebanese American, Amin Rihani, Holmes and his consortium of mainly British backers obtained the concession, only to let it lapse a few years later by failing to keep up the annual payments or to find traces of oil. Undaunted, Holmes pursued his dealings on the neighbouring island of Bahrain in 1925, selling his concession on to a subsidiary of the Gulf Refining Company two years later. The geological evidence here was good, although Anglo-Persian was extremely doubtful that oil would ever be found either there or in the Arabian peninsula. The trouble was that Bahrain came within Gulbenkian's Red Line. The Iraq Petroleum Company insisted that Gulf, still a member of the IPC, obey the all-or-nothing rules of the association. Gulf offered to sell its concession to the consortium. It wouldn't take it but none the less insisted that Gulf couldn't own it alone either.

Gulf sold on the concession to Socal, which was seeking foreign expansion, and turned its attention instead to Kuwait. Holmes, who had been active there since 1922, was instructed in 1928 to proceed to the sheikhdom to firm up a concession there. With a carefully considered campaign of bribery and flattery, noted down in detail for reimbursement by Gulf, Holmes proceeded steadily enough. 'It is interesting to observe,' he remarked at the time, 'the greedy attitude of the Sheikh of Kuwait exhibited when I hinted to him that the Company I was representing had an American tang about it.'[2] Holmes soon came across precisely the same difficulties as he had in Saudi Arabia. Hearing of his endeavours, the British moved in to remind the Sheikh of previous treaties, which promised British precedence in commercial exploitation, and they brought in Anglo-Persian, in the shape of Archie Chisholm, a young, languid but shrewd relation of the new chairman Lord Cadman, to counter Gulf's offer.

Anglo-Persian was not necessarily interested in developing oil in Kuwait itself but it didn't want anyone else to come along and develop alternative supplies either. 'Although the geological information we possess at present does not indicate that there is much hope of finding oil in Bahrain or Kuwait,' went a report to the directors of the company in 1924, when the rivalry between Holmes' consortium and Anglo-Persian for the concession had first started, 'we are, I take it, all agreed that even if the chances be 100 to 1 we should pursue it, rather than let others come into the Persian Gulf and cause difficulties of one kind or another.'[3]

Anglo-Persian followed this directive by leaning on a not-always admiring Colonial Service to back its claims in Bahrain, Saudi Arabia, Kuwait and elsewhere in the Gulf. Holmes proceeded by claiming equal right to British official support as the representative of a London-based consortium and also established himself locally by drilling wells for water, thus gaining geological knowledge as he went along as well as building up his friendships and contacts.

'He was one of the most remarkable men that I have met,' recalls Archie Chisholm, 'the real genuine bedouin, as they used to say, dyed-in-the-wool British businessman; a very rough diamond, a great believer in the personal touch, a great believer in playing the Arabs at their own game, while we thought we would play the Arabs at our own game.'

The two rivals clashed head on in 1931, much to the delight of the Sheikh who could now play them off against each other. 'We were hot adversaries,' Chisholm remembers. 'We were pulling every trick out of the bag to try to do each other in and at the same time were living as neighbours in Kuwait, which in those days was a small walled Arab town.' The British played it their way, using the British connection to the maximum advantage and hinting at a withdrawal of British military protection of the Sheikhdom if they were turned down. The Americans responded by appealing to the US State Department at the breach of the 'open-door' policy and used the American Ambassador to London, then conveniently Andrew Mellon, Gulf's biggest shareholder, to press the case through Whitehall. Holmes dealt and intrigued 'the Arab way'. Chisholm was equally insistent, turning on the full force of his Anglo-Irish charm. They matched stratagems in the morning at the palace and exchanged pleasantries in the evening at the social club or after church on Sundays. Eventually, as Chisholm comments, 'Our respective chairmen decided that the only people who were benefiting out of this competition between Gulf and ourselves to get the Kuwait oil concession was the Sheikh, because we were bidding each other up and he was sitting like a pig in the middle watching the price improve.'

It was finally settled in a joint concession, agreed on Christmas Eve, 1934, largely because of Anglo-Persian's growing concern at the oversupply of oil at a time when the world demand had been badly hit by the Depression. This concern was reflected in an address made by Lord Cadman to an audience of American oilmen in November 1932: 'Consumption everywhere decreased, competition is too keen; and prices contain no margin for gain. One possible step toward rehabilitation of trade is evidently the readjustment of supply to demand and the prevention of excessive competition by allotting to each country a quota which it will undertake not to exceed.'[4]

The result, for the Sheikh of Kuwait, was a 75-year concession agreement on much less attractive financial terms than he had been led to expect from the individual offers made earlier on and with lower royalties than most of his neighbours had obtained. In addition, a clause giving him the right to appoint a director to the board of the local company was deleted. More significantly, Anglo-Persian persuaded Gulf to sign an agreement early the following year ensuring that any development of Kuwaiti oil would not 'upset or injure' the world marketing position of either company. Should Gulf require more oil than Anglo-Persian wanted to produce in Kuwait, Anglo-Persian, as the 'sole judge' of the situation, was to be allowed to provide alternative supplies from Iran and Iraq.

Frank Holmes himself and his backers gained little from his endeavours. Unlike Calouste Gulbenkian, 'Mr Five Per Cent', he sold out his royalty rights in Kuwait for a few thousand pounds. None the less, he had started, and helped bring competition to the great Arabian oil adventure: in Bahrain, where oil was found on the first well in 1932; in Kuwait, where after a slow start to drilling, one of the largest fields in the world was discovered at Burgan in 1938; and above all in Saudi Arabia.

The Last of the Warrior Princes

The developer this time was not a member of the IPC consortium at all but Socal, to whom Holmes and Gulf had sold out in Bahrain. The intermediary was Henry St John Philby, adventurer, writer, father of the British spy Kim Philby and a convinced supporter of the Arab nationalist cause. The ruler was King Abdul Aziz Ibn Abdul Rahman, known as Ibn Saud, 'son of the house of Saud'. The last of the Middle East's warrior princes, a man of huge physical presence with a passion for close encounters both on the battlefield and in the bedchamber, Ibn Saud had carved his way to a kingdom through conquest, using battle, bribery and marriage to unite the disparate tribes. Philby recorded him as confessing when quite an old man 'to having married no fewer than 135 virgins, to say nothing of "about a hundred" others during his life, though he had come to a decision to limit himself in future to two new wives a year, which of course meant discarding two of his existing team at any time to make room for them.'[5] He was equally boastful of his prowess with the sword. The two were parallel routes to power in the vast and empty land of Saudi Arabia over which he finally proclaimed himself King in 1932.

Although paid an annual retainer of £60,000 by the British in recognition of his support against the Turks during the First World War, when he had conquered the holy city of Mecca, Ibn Saud's road to expansion caused too many problems with neighbouring rulers for the British ever to feel comfortable with him. He, for his part, was well enough aware of Britain's failure to meet its promises of Arab independence after the First World War to trust the British either. It was partly because of this that Philby, who, like T. E. Lawrence, felt strongly about the British perfidy, was drawn to Ibn Saud and the King to him.

When Socal turned their eye to nearby Saudi Arabia in 1932,

therefore, it was natural enough that they should seek the good offices of Philby. Philby, although working as an agent for the Ford Motor Company, was badly in need of money. So was the King, despite his suspicion of the infidel, a suspicion that Holmes' failure to keep up with the rents on his concession had done little to mollify.

The Depression had cut the flow of pilgrims to Mecca and Medina on which the King depended for much of his income. Out on a walk with the King in the autumn of 1932 Philby, struck by the air of despondence in his patron, asked him why he was so gloomy. As Philby recalled:

> He sighed wearily and admitted that the financial situation of the country was seriously worrying him. . . . I replied, as cheerfully as possible in the circumstances, that he and his government were like folk asleep on the site of buried treasure, but too lazy or too frightened to dig in search of it. Challenged to make my meaning clearer, I said I had no doubt whatever that his enormous country contained rich mineral resources, though they were of little use to him or anyone else in the bowels of the earth. Their existence could only be proved by expert prospecting.

To this plea to bring in foreign interests, the King replied 'almost beseechingly, "Oh, Philby, if anyone would offer me a million pounds, I would give him all the concessions he wanted." '[6]

The offer of £1 million was not forthcoming but Socal did promise Philby a retainer to advance their cause, which he accepted while also encouraging the Iraq Petroleum Company to make a counter bid. It was like Kuwait. The two companies, with Philby constantly moving between them, upped the bids until it came down to a question of hard cash. The Iraq Petroleum Company finally failed because of its reluctance to offer sterling instead of rupees (the currency used in the Gulf) and their refusal, because of British government controls, to offer the money in gold. Socal offered £50,000 in gold. With a final exclamation of, 'Very well! Put your trust in God and sign', the King cut through the detailed talks and ordered his Finance Minister to accept their offer in May 1933. Ignoring the new American regulations controlling the export of gold, and the American Government's refusal to sanction Socal's request, the company bought 35,000 gold sovereigns on the London market and shipped the money out to Jedda within days of the expiry of the three-month deadline for the first payment.

Over the next half century Saudi Arabia was to prove the most important of all the oil kingdoms in the Middle East, although it took seven wells, drilled in searing heat and appalling conditions before oil was finally found there in 1938. For the immediate future it proved a decisive counterbalance not only to the weight of the partners in the Iraq consortium but also to British influence in the region. Saudi Arabia became firmly American, placed in the hands of a newcomer to the international market which had few outlets of its own.

The End of the Red Line Agreement

The danger to the existing Middle East concessionaires of this new presence in the region was clear and almost as soon as Socal discovered oil in Bahrain, all the IPC partners joined forces. Their efforts to buy Socal's Bahrain and Saudi concessions were firmly resisted and attempts to modify the Red Line Agreement so as to exclude the two countries and allow individual companies to do their own deals failed because of divisions among themselves and the firm opposition of Gulbenkian and the French to any change in the status quo. Socal, however, needed markets in which to sell its oil and funds to develop its transportation facilities. It found its own solution in a company outside the IPC – Texaco. Under the resulting union of Caltex, formed in 1936, Texaco took a half interest in Socal's concessions in the Gulf. In return Socal received a half interest in Texaco's refinery and marketing facilities east of Suez with an option to include its European operations later. Caltex created a new and powerful challenge to Standard and Mobil with their Stanvac union and to Shell and Anglo-Iranian. It also brought Socal into the fold of the Seven Sisters. Like the rest, she now had an interest in marketing stability and firm prices.

The Red Line Agreement as such did not long survive the Second World War. The war brought its share of problems to Saudi Arabia as well as to Iran and Iraq. Pilgrimage traffic collapsed, oil development was held up for lack of materials and equipment and King Ibn Saud once again found himself short of money. In 1941 he demanded $6 million in advance royalties. The Socal-Texaco company felt able to offer only half that sum and turned to President Roosevelt for help in making up the numbers.

Into the discussions, therefore, came Harold Ickes, President Roosevelt's spiky Secretary for the Interior and the man in charge of organizing the industry during the war. Ickes, who had been intimately involved with Roosevelt's plans for National Reconstruction and charged with developing the oil industry's role in it, saw the opportunity for the State to gain the same kind of direct access to Middle East reserves as Churchill had obtained for Britain in the First World War. He suggested to shocked Texaco and Socal executives that the US Petroleum Reserves Corporation buy out their joint company in Arabia – a proposal which, on his account, had them 'nearly falling off their chairs'.[7] When this proved too much for them to swallow he suggested a government share of 70 per cent, later reduced to 51 per cent and finally to a third. The whole scheme caused such a furore within the American oil industry, which saw it as the first insidious signs of nationalization, that he was forced to withdraw the scheme altogether, much to the companies' relief.

The alternative he proposed was even more ambitious but it avoided direct partnership. He suggested building a government-financed pipeline to take oil from Saudi Arabia and Kuwait across the Arabian peninsula and up to the Mediterranean. The plan, in his eyes, would bring the Americans firmly into the Middle East arena as equal partners with the British. While the idea naturally proved welcome to Texaco and Socal, it was rather less than welcome to the home oil industry, which was terrified that the pipeline would be used to flood the United States with cheap imports. (Indeed one of Ickes' ideas was that, in return for financing the line, the Government would gain a quarter of its through-put at reduced prices.) So this proposal was dropped too, 'done to death without benefit of clergy' in Ickes' words, thus putting an end to the United States' brief flirtation with state intervention, at least until the 1973 energy crisis. The line was eventually built by the Aramco partners in Saudi Arabia as the Trans-Arabian Pipeline (Tapline) in 1949, without state aid.

None the less, it marked American determination to ensure that Saudi Arabia remained under their influence. Money was made available to the King on the clear understanding that it was paying off British advances, which had only been temporary. Towards the end of the war, King Ibn Saud made his first trip ever outside Arabia to meet President Roosevelt and Winston Churchill in Cairo in February 1945, both on their way home from the Yalta Conference with Stalin. Significantly, the King travelled on an American destroyer, complete with a large Arab-style tent on deck to accommodate himself and his retinue of forty-eight guards, slaves and servants. His talks with Roosevelt, aboard the American cruiser *Quincy* went well. He admired the President's wheelchair and was given his spare one. He asked for support for the Palestinian cause and got a promise, ignored by later administrations, that the United States would not develop any new policy without consultation. The talks with Churchill at the Hotel du Lac at Fayoum Oasis were less easy. Churchill insisted on smoking and drinking during the meal and could make no promises on the Palestinian question. As at Yalta, the force of his personality masked the obvious beginnings of the decline in British influence in the post-war world.[8]

The growing force of American commercial might was to be seen almost immediately afterwards in Saudi Arabia when the Texaco-Socal partnership, or Aramco (Arabian-American Oil Company) as it was now called, looked to other American partners to help finance the development of the Mediterranean pipeline scheme. The partners they sought were Esso and Mobil. Both wished to expand their markets in the post-war world but both were restricted by their membership of the Iraq Petroleum Company. The obvious answer was to break up the Red Line Agreement. Shell was willing. It too wanted the freedom to develop its own resources in the Middle East. Anglo-Iranian was more dubious but was partially persuaded by a long-term deal to sell oil to Esso. The objectors of course were CFP of France and Gulbenkian. Their share had been seized as 'enemy property' during the war and they stood to lose by any such deal. Mobil and Esso announced that they were buying into Aramco at 30 per cent and 10 per cent respectively. Gulbenkian promptly announced his intention of taking legal action to insist that he too be offered a share under the IPC agreement of 1928.

Negotiations dragged on for a year and a half. As a sop the French were offered the return of revenues taken over during the war. Gulbenkian, denying that his share should ever have been taken in the first place since he travelled under an Iranian passport, held out. He believed, as his son Nubar informed the companies, that 'the time when others were pressing you for an urgent decision is the time to take it slowly'. The date fixed for the start of Gulbenkian's legal proceedings in London drew nearer and in desperation with this deadline hanging over them, a troop of top executives from the Iraq consortium companies flew out to his hotel in Lisbon in November 1948. As the minutes to the London court action ticked by, they finally reached an agreement at two o'clock in the morning, too late to obtain champagne from the hotel kitchens. Cheap wine at an all-night café had to suffice to toast the ending of the Red Line Agreement. The dominant constraint on company action in the Middle East had at last been removed.[9]

The Legacy of Achnacarry

The decisions of November 1948 did not, however, end co-operation between the companies. They were now tied more closely than ever to an interlocking series of consortia and crude-oil supply arrangements through the Gulf. They were also held by both the spirit and the letter of the Achnacarry Agreement which had established a worldwide structure of oil prices based on the price in the Gulf of Mexico. Continuous pressure from the British and American governments in the years after the war modified the system, as discussed in Chapter 3, but the modifications were all made in the direction of reducing the formal export prices of the Middle East in comparison with American oil.

That the fate of the price of their most important export should be moved about in this fashion was bound to make the oil producers of the Middle East increasingly irritated, whatever the commercial logic that may have been behind it. The Second World War had, like the First, brought with it a resurgence of nationalism through the Middle East, intensified by the Palestinian troubles and the creation of the State of Israel in 1948. Britain was caught up in the middle and blamed by all sides, despite its tendency to sympathize with the Arab cause. Truman overturned Roosevelt's earlier promise of American–Arab consultation with the argument that he had 'hundreds of thousands of people who are anxious for the success of Zionism. I do not,' he continued, 'have hundreds of thousands of Arabs among my constituents.'[10]

The New Nationalism

The nationalist cause found a ready target in the oil industry as a symbol of Western domination and Arab humiliation. The terms of the concessions gave most of the concessionaires virtually unlimited rights to huge areas for periods of between seventy-five and ninety-nine years. Companies could arrange the production and off-take and determine the price without any consultation with local governments, still less with any representation on the boards of the Middle East subsidiaries of the internationals. It may have seemed fair enough at the time the concessions were granted, during the late twenties and early thirties when oil was in surplus supply, the Depression was causing a dramatic fall in prices and the Middle East seemed (and was) a tiny producer in a world dominated by American oil. It did not seem so acceptable to a post-war world in which the United States had ceased to export and virtually all the growth in the industry was coming from the Middle East.

It is easy to see that the older Arab generation, brought up under the shadow of British political 'advisers' and always desperately in need of the cash oil brought them, might have accepted the status quo, particularly since rising production output at least brought them rising incomes. Where Ibn Saud was paid $½ million a year for his oil in 1938 (and the money was paid directly into his privy purse in the manner of medieval monarchies), by 1949 this had grown to $39 million. Between the end of the war and 1950, royalty payments to Kuwait increased from $800,000 to $12·4 million; Iraqi payments rose from £2·6 to £6·7 million and Iran's from £5·6 to £16 million.

By then, however, the expectations of the producing governments had been radically altered by Venezuela's success in forcing the oil companies to accept the principle of a 50/50 profit sharing on oil. Venezuela, like Mexico a generation earlier, had only recently emerged from a prolonged period of dictatorship under General Gomez, who died in 1936. Gomez had single-mindedly encouraged the oil companies to develop the country to become the United States' biggest supplier and his country's largest source of income. Emboldened by the United States' dependence on Venezuelan supplies during the Second World War, his successors sought to increase that income greatly in order to fund wider social development programmes. In 1943 the military regime of General Angarita negotiated a new agreement with the companies giving the state an 80 per cent increase in revenue. Immediately following the war, the opposition Acción Democrática was elected on a platform of increasing this revenue by a special tax that would give Venezuela 50 per cent of the export price for its oil. Re-elected to power in 1948, the party finally got the oil companies to accept, just before the politicians were ousted by a military coup.

The 50/50 agreement applied specifically to Venezuela and was achieved by a complicated formula for Venezuelan income tax laws. But it was a simple rallying cry which was soon taken up on the other side of the Atlantic. Middle East revenues at that time were based on a straight sum, a 'royalty' per tonne of export. In most countries they were relieved of local income or sales taxes. Led by Ibn Saud, the Gulf countries now demanded a similar rate of income to Venezuela. The oil companies were, at first, most reluctant to consider moving towards this increasing of royalties to half the selling price of oil. After protracted negotiations and discussions with the American Government, however, the Aramco consortium agreed in 1950 to a mutually beneficial deal with the Saudis to pay 50 per cent income tax on their production, together with a 12·5 per cent royalty on the oil. The tax and the royalty were henceforth to be based on the official selling price, or 'posted price' of the oil. Income tax was to be paid after royalties had been deducted, so reducing its effective rate.

The beauty of this arrangement was that foreign income tax, unlike royalties, could be set off against American domestic taxes by the corporations, so that the huge profits they earned from their Middle East operations remained tax free in their home base, a concession granted to BP, Shell and other British companies by the British Government not long after. The American Government accepted it when negotiated partly out of consistency with the general principle that companies should not be taxed twice on the same operation. A more cogent factor in its agreement to such a loophole was its desire to help the Arab governments, and the Saudis in particular. 'The State Department,' asserts Jerome Levinson, chief counsel to the US Senate Foreign Relations Committee, whose hearings into the multinational oil corporations uncovered the issue in 1974, 'thought that it was in the US interest to provide funds for the Saudi Arabians and the sheikhdoms for political stability through the foreign tax credit, through the companies, so that they would not have to come to Congress for annual aid appropriations, and thus get into the Arab-Israeli question. So a decision was made for foreign policy reasons to try and provide revenues for stability in the Persian Gulf.'

To the Senate Foreign Relations Committee, chaired by Senator Frank Church, the revelation was a devastating illustration of the way in which the American Government had 'delegated' a key foreign policy decision to the oil companies. For the Aramco partners at the time, and the other companies involved in the rest of the Middle East which quickly followed on the agreement, it was a convenience. It meant that the industry could concentrate its profits on the production end of the business, thus avoiding consumer tax. For the producing governments it meant dramatically increased revenues. By the mid-fifties the oil companies were paying 80 cents a barrel to the host governments as against an average of 22 cents in 1950. The difficulty for the future was that it made the companies' official price postings into the tax-reference price for oil. If the companies ever wanted to reduce those postings, the producers would be bound to resist.

For the younger Arab generation, on the other hand, it wasn't just a question of money. It was a question of control and pride. The concession terms, which seemed to place the exploitation of their key resources entirely in the hands of faceless executives meeting thousands of miles away, were particularly irksome. Oil became the symbol not only of a hated West but also of a growing decadence at home. As the newly educated students watched the deserts bloom not with industrial and agricultural projects but with vast and ever more magnificent palaces of the scions of the royal families, inevitably oil became associated with the propping up of corrupt and authoritarian regimes.

When General Abdul Nasser swept to power in Egypt in 1954 and then, slapped down by the withdrawal of Western aid for his treasured Aswan Dam project, he nationalized the Suez Canal in 1956, it seemed as if the Arabs had at last found a voice. Nasser lost the battle. The British and French, in secret accord with the Israelis, invaded and were only finally forced out by American and UN pressure. Nasser had won the war of words, however, becoming the hero of an entire generation of Arabs. It was the Allies, particularly the British and Americans, who ended up divided. The days of the monarchy in the Middle East seemed numbered, all the more so when the Hashemite monarchy was overthrown by a coup in Iraq in 1958.

The oil companies were slow to accept the new mood. Brought up in the days when commercial profit depended on the terms that could be extracted from personal negotiations with – and bribery of – the local ruler, they were reluctant to move from the world of the palace to the world of the populace. All along, complains Howard Page, the man brought in by Esso to negotiate the companies out of tight corners in Iraq in 1952 and Iran in 1954, they remained fixated by the concept of sanctity of contract, afraid that any concession on their part in one country would be used as the excuse for even tougher demands from others. They thus obstinately refused to agree to the demands of the new Iraqi Government in 1958 for greater national control of the industry only to find that two years later a further revolution brought in a new regime that simply deprived the consortium of the undeveloped area of the concession (some 97 per cent). Demands for state participation coming from the new Western-educated technocrats of the ministries of petroleum in Kuwait, Saudi Arabia and Iran were persistently ignored.

'The older members of the senior hierarchies in the different companies,' remembers Sir David Barran, then something of a young Turk in the ranks of Shell who later became its chairman, 'felt that it was a mistake to take the representatives of these producing governments into your confidence, to open up the books if you will, and disclose the facts. It was much better to leave them to stumble along to find out for themselves. In that way it would take longer.'

By their own lights, and the lights of a time when oil was in surplus and the producing countries were competing for volume, the companies had a point. In the short term at least the monarchical regimes of the Gulf actually survived far better than anyone had expected. Riots in Kuwait at the time of the Suez invasion and religious objections to westernization in Saudi Arabia forced some distancing from and a measure of independence between rulers and companies. The Shah of Iran, returned to power after Mossadeq's demise, was determined to show that he could hold his own with the newly formed consortium and get as much as the next man.

But the Gulf, where the oil reserves were, was not the Levant. Nasser's dreams of unity with Syria and Iraq ended in friction and disillusionment whilst his intervention in the anti-British revolt in Aden served only to frighten the rulers of the Gulf and drive them closer to the West. As far as the companies were concerned unilateral seizure of the undeveloped concession area of Iraq, 'Law 80' as it became known, and the threats of legal action which followed it, were preferable to their giving in to the concept of a country's right to order how much investment they were to make and how much oil they were to produce.

There was no shortage of oil in the world. The oil companies managed, partly because of their ability to adjust supplies among themselves, to get over the Iranian problems of 1954, the temporary closure of the Suez Canal in 1956 (and its much longer closure after the 1967 Six-Day War). Law 80 led to a drastic slowdown in development in Iraq that was not entirely unwelcome to the consortium members, all of whom had alternative supplies elsewhere. Their main preoccupation was to balance out the demands of the various countries for greater 'liftings', or exports, and prevent them from flooding the market.

The Producers Unite

What finally drove the oil producers together was not the success of the majors in controlling the market but their failure. The resolution of the Suez crisis and the reopening of the canal and the end of the Korean War had seen prices fall again. The rates for hiring tankers

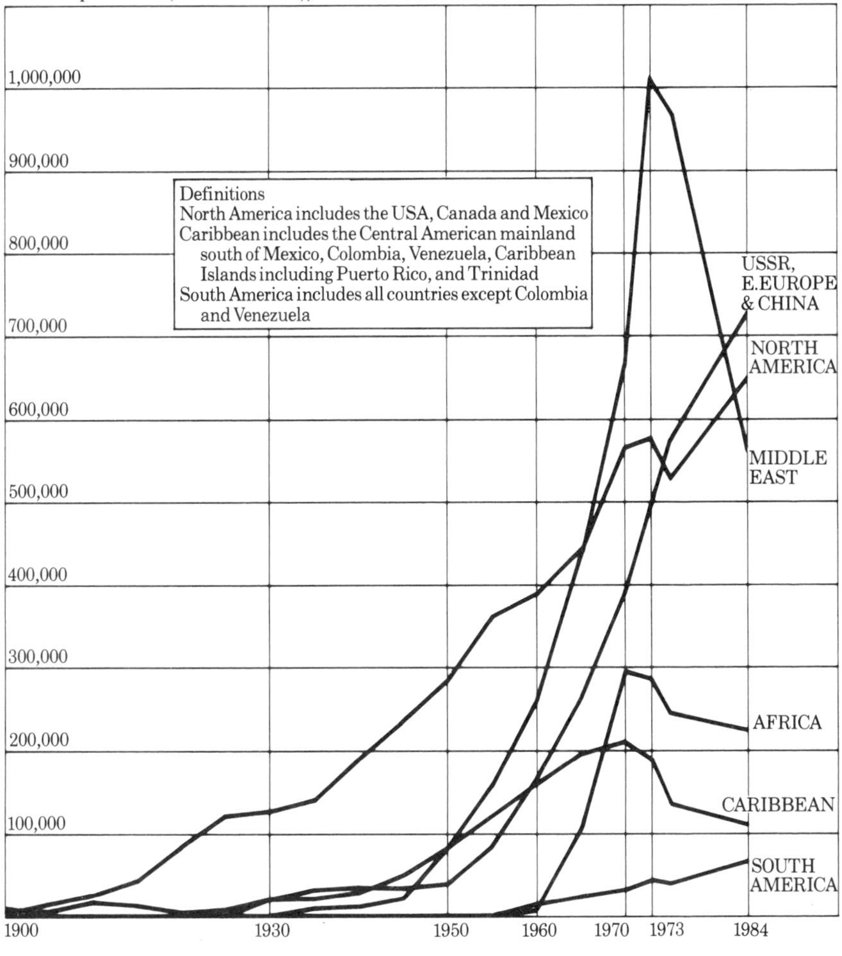

Year	NORTH AMERICA	CARIBBEAN	SOUTH AMERICA	USSR, E.EUROPE & CHINA	AFRICA	MIDDLE EAST
1900	8,743	-	38	11,128	-	-
1930	129,548	24,424	3,171	25,307	278	6,143
1950	285,198	87,559	5,800	42,834	2,380	86,076
1960	390,581	162,569	17,452	167,200	13,790	261,836
1970	568,189	212,124	34,859	390,600	298,875	687,601
1973	578,006	193,613	47,536	493,101	289,508	1,046,710
1984	651,738	113,270	70,348	722,825	228,419	563,308

Source: British Petroleum

The sources of world oil during the twentieth century. Until 1950 more than half the total production was accounted for by North America. By 1960 it is the Middle East which dominates the picture, only to fall back dramatically after the first oil shock of 1973.

dropped and the oil companies sought a fall in posted prices to reflect this. BP, keen to rebuild its position after the Iranian settlement and to establish a stronger hold on the markets, followed a cut of 10 cents in the price of American oil with an 18-cent cut in Middle East postings in 1959, a move taken up in its turn by the other producers. The price change effectively broke the last connection between Middle East and Venezuelan prices which had existed since the Achnacarry Agreement and in doing so threatened a price war between Venezuela and the Middle East. The situation became all the more tense when, a few months later, the United States introduced mandatory controls on exports. Venezuela, whose oil policy was now in the hands of Dr Juan Pablo Perez Alfonso, was seriously worried.

Alfonso, who had been an oil official in the post-war Venezuelan Government and had helped formulate the 50/50 profit-split campaign, was a man of personal simplicity and total dedication to the national cause. He sought both a special agreement with Washington guaranteeing Venezuelan oil access to the American market and co-operation with the Middle East to fight together against the price cuts. His first tack was unsuccessful. Washington refused. His second found a much readier hearing.

The Oil Minister of Saudi Arabia was Abdullah Al-Tariki, a man with whom Alfonso had a natural empathy. 'A real fire brand', as Howard Page of Esso remembers him at the time, Tariki had been embittered by his early experience of the United States as a student. 'He came to New York,' according to Page, 'and being dark he couldn't get into any hotel room, except the scruffiest one. . . . Then he went down to Texas, where he was taken to be a Mexican, which in those days was considered to be lower than a black, in Texas anyway. He wasn't accepted at all as a person and of course he felt that very deeply.'

Both Tariki and Alfonso, according to Dr Alirio Parra, then a colleague of Alfonso's, had seen the workings of the Texas Railroad Commission which had been set up before the Second World War to establish an orderly management of production in the American southwest. They had become convinced that not only should a similar approach be adopted worldwide but also that it was essential to push for state involvement and ultimately control in the operations of the oil companies. 'It was these basic philosophical considerations plus the circumstance of their having both lived in the United States, of their having met and discussed things in the US as well as in Venezuela and in the Middle East', comments Dr Parra, 'that really heightened the bond between them and gave them both a sense of purpose and a sense of direction in getting across what must have been to many observers at that time an impossible task.'

The first round of BP-led posted-price reductions provided the impetus for action. Using the opportunity of a meeting in Cairo of the newly formed Arab Petroleum Congress, attended by Venezuela as an observer, Venezuela, Iran, Saudi Arabia, Kuwait and the United Arab Republic (Egypt and Syria) held a series of secret talks at the yacht club. The upshot was an informal understanding, or gentlemen's agreement, as it became known, to co-ordinate their oil policies, seek common tax terms and take a much more leading role in the oil business, rather than letting themselves be dictated to by the companies.

On a stage set for the major producers to move towards some sort of association, the oil companies dropped a new bombshell in the shape of a second round of posted-price increases in the summer of 1960, this time led by Esso, and implemented without any prior consultation with the producer governments. Howard Page argued strongly against such action at the time. He remembers warning his colleagues that: 'all hell would break loose' if we cut the price in all those countries out there . . . without waiting to see if we could negotiate a reasonable settlement with them, which I felt was possible to do.' Esso, however, thought that the posted price should reflect the market price. And market prices were dropping, partly because the Russians, encouraged by Mattei, had restarted exports and were undercutting the existing suppliers to Europe and India.

The result was, as Page had predicted, electric. Not all the companies followed Esso's lead. Esso, supported by Shell, Texaco and Socal sought a 14-cent cut. BP, fearful of a repetition of the fuss caused by the previous year's move, countered with a more limited cut of 10 cents and was supported by Mobil, Gulf and CFP. Esso then added to the confusion by drawing back at the last minute from introducing a similar cut in Venezuela, where the concession terms required prior consultation with the Government on any price change.

The Founding of OPEC

Faced with the prospect of a dramatic fall in their incomes and a divided oil industry, the producers set aside their own differences to answer Tariki and Alfonso's call to arms. In the words of Dr Parra 'I was sitting in front of Perez Alfonso at his desk on this September afternoon in 1960 discussing prices and production levels when his secretary walked in with a telegram. Perez Alfonso opened the telegram, read it and all of a sudden I saw him stand up, wave the telegram in his hands and he said: "We've achieved it, we are on the way to forming an international oil agreement."

'The telegram came from Baghdad from the then Minister of Oil, inviting him to meet in Baghdad on 11 September, with a view to setting up a commission representing the producing countries in the Middle East and Venezuela.'

OPEC was founded with just five members: Venezuela, Saudi Arabia, Iran, Iraq and Kuwait, the 'Big Five' oil producers of the time. Qatar joined the following year. Libya and Indonesia came in in 1962 and by the time of the 1973 oil shock the organization numbered thirteen with the further additions of Abu Dhabi and Dubai (both merged into the United Arab Emirates), Algeria, Gabon, Ecuador, and Nigeria. With determination the founding members declared their intention of, first, stabilizing the price of oil and insisting that no future changes be made without consultation with the host government; second, looking at ways of achieving assured income and stable prices through, among other things, the regulation of production; and third, supporting each other should the oil companies try to pick them off one by one. 'We have formed a very exclusive club,' Perez Alfonso declared enthusiastically. 'Between us we control 90 per cent of crude exports to world markets and we are now united. We are making history.'[11]

History was not made quite as radically as Alfonso or Tariki had perhaps hoped. Certainly, the threat of united opposition was enough to force the oil companies to readjust the price cuts to a common 10 cents but not, as OPEC had wished, to withdraw them altogether. The organization, which first set up shop in Geneva, discovered a few years later that it had failed to gain the requisite Swiss approval for the establishment of an international governmental body on its soil. Reluctantly it was forced to switch offices to Vienna. The oil companies refused for a long time to treat with OPEC as a single body, preferring to continue bilateral arrangements with individual governments. The member states, for their part, failed to pursue the all-important issue of pro-rationing their oil, despite Alfonso's constant urgings. There simply wasn't the agreement. While Venezuela, with its limited reserves and higher production costs, would obviously benefit from a pact which would apportion everyone's production at a set profit, the Middle East states still saw it as in their interests to maximize revenues by increasing volumes. The Shah of Iran in particular, was determined to restore his country's place as the premier producer in the Middle East after the upheavals caused by Mossadeq.

Tariki was replaced as Oil Minister in 1962 by Sheikh Yamani, a lawyer where Tariki was a geologist. A gradualist in outlook, Yamani was just as keen on national participation but more willing to work through the system as it was, in partnership with the companies rather than in direct opposition to them, and was less committed to the concept of production regulation. Outside OPEC at the time were a number of smaller producers such as Dubai and Abu Dhabi with every incentive to get their youthful oil industries producing to the maximum as quickly as possible rather than constraining production within overall targets set by countries which had already achieved relatively high outputs. Instead of regulation, the oil producers concentrated their attention on tax levels and the technical, although highly lucrative, question of how royalties were to be treated in the tax mix imposed on the companies. The 'charter for change', which OPEC at first seemed to promise, soon became more a charter for defence, a charter for the status quo. Prices in the market fell progressively as surpluses built up with new oil from Nigeria and North Africa. Never again, however, did the oil companies dare to bring down the posted, or tax-reference, price in line with this fall. The oil producers' income was guaranteed so long as demand grew and their own volumes kept up.

The Libyan Storm

In the end the break came not from OPEC or even the Middle East but from the new producers of North Africa and the most traditionally conservative of them all. Libya, which had originally offered highly attractive tax and concession terms in order to encourage companies to come in and search for oil, had been pressing for sometime for a rise in posting to reflect its geographical advantage of closeness to the markets. That advantage had been greatly increased by the June Six-Day War of 1967 which had seen the Israelis move across the Sinai Desert to take up positions along the Suez Canal, thus shutting it off completely to oil or any other traffic. Freight rates had risen but Libyan postings had not moved to anything like the same extent when the regime of the King Idris was suddenly swept aside by a revolt orchestrated by a group of young army officers on 1 September 1969. Their leader was Lieutenant Muammar al-Gadaffi. Mohammed Heykal, editor of *Al Ahram* and a close confidant of President Nasser's in Egypt, remembers Gadaffi as a 27-year-old of quite extraordinary innocence and purity, charged with Arab nationalist zeal.

That innocence and zeal, having rid the country of its British and American bases, was now directed towards the oil companies. Gadaffi demanded, as of immediate right, a rise in postings of between 40 and 50 cents per barrel as well as a retroactive payment to make up for the underpricing in the past. He threatened outright nationalization if the companies did not accede. 'People who lived without oil for 5000 years,' said Gadaffi, referring to the Libyans when addressing all twenty-one companies operating in his country at the beginning of 1970, 'can live without it again for a few years in order to attain their legitimate rights.'

The oil companies, led by Esso, the largest producer of all, at first took this simply as a negotiating ploy. Gadaffi, now promoted to colonel, meant it. He also had the advantage of coincidental timing. In the first place, he was able to combine his approach with that of Algeria and Iraq, both of whom were in dispute with their respective oil concessionaires. The Algerians were negotiating a new deal to replace their five-year agreement with France. Iraq had undergone yet another revolution, this time resulting in the triumph of the Ba'athist party. All three countries had chosen to make extensive new demands on the oil companies just as the market was turning sharply in favour of the 'short-haul' producers of the Mediterranean – Iraq by virtue of its pipeline to the Mediterranean, and Libya and Algeria by virtue of their position in North Africa. An accidental break in the Tapline taking Saudi oil through Syria to the Mediterranean tightened the squeeze and the free-market price for short-haul oils rose over 50 cents. Suddenly the oil industry was going to find it difficult to do without Libyan oil.

Gadaffi, with the help of a shrewd deputy, Major Abdul Salaam Jalloud, increased the stranglehold with ruthless skill. On the grounds of 'conservation' – grounds that had some force in reality considering the rate at which some of Libya's fields were being depleted – Libya first cut back production allowables progressively, reducing her overall output from 3·7 million barrels a day to 2·9 million over a period of six months. The President and his deputy then concentrated their attention on two companies, Esso as the largest producer and Occidental as the company most dependent on Libya for its international supplies. Esso, with plenty of alternative sources at its disposal played it coolly. Occidental, however, threatened with nationalization and the loss of its one major source of oil outside the Americas, was in a particularly vulnerable position.

Dr Armand Hammer, its septuagenarian chairman, a man of limitless daring and shrewd negotiating sense, had taken a small and failing Californian oil company and made it into one of the ten biggest in the world. He had gained concessions in Libya by the imaginative expedient of offering to find water in the desert and build chemical plants to make fertilizers if he was awarded territory. His offer, he recalls, was tied with ribbon in the Libyan national colours to give it that extra *je ne sais quoi.*

As soon as it became clear that he was being singled out along with Esso, he flew to New York to see Esso's chairman, Ken Jamieson. Esso was no lover of Hammer, ever since he had quickly moved in to take over a concession of theirs after it had been nationalized in Peru. Perhaps not surprisingly when Hammer asked whether Esso would provide him with oil were Oxy to be nationalized, the reply was blunt: only at full commercial prices. A phone call to Sir David Barran at Shell in the middle of the night brought a more sympathetic response but nothing more concrete. Shell, very worried by the level of Libyan demands and particularly their insistence on retroactive payments, was ready to offer crude oil of its own but couldn't organize other companies to do so, particularly when it might involve oil from non-Libyan sources and raise problems of anti-trust.

So when Dr Hammer finally received the expected call summoning him to Libya, he was well aware that he had no back-up in the event of nationalization. He also went in some fear of his life, for he was known to have been a close friend of the deposed King's. Instead of being met by a firing squad on his arrival he was warmly greeted by Major Jalloud as the first of the oil company bosses to come in person. 'We sat down at a table,' says Hammer, 'and he unbuckled his belt and put his revolver on the table while we started negotiating. It wasn't a very good sign to me but nevertheless I held my own with him and found that what he really wanted was an increase in taxes of 30 cents a barrel. It seemed little enough to me and I readily agreed to it.'

Occidental's cave-in made it extremely difficult for others to refuse the same terms. Even so, Shell and BP, as the two biggest non-American producers there and therefore not subject to anti-trust, made a strong effort to get at least the majors to stand firm. Sir David Barran flew to New York with Sir Eric Drake of BP, lunched with the British Foreign Secretary, Sir Alec Douglas-Home at the United Nations and then went on to the State Department in Washington to meet with other companies. The State Department, influenced partly by their oil counsellor, Jim Akins, a convinced Arabist who sympathized with the Libyan demands, was not receptive. 'They had just been successful,' remembers Sir David, 'in mounting a plan to get rid of the Palestinian refugees from northern Jordan and they were so full of their success on that, they thought that they had got the whole Middle East problem settled and there weren't any other Middle East problems worth bothering with. So we never really got to first base with them. And from then on it was perfectly clear to me that nobody in Libya was really going to stand up and be counted in a common line and we were all just going to be picked off one by one.' So they were. Texaco proved to be the first of the majors to fall in with the Libyan demands. The others followed with the exception of Shell. When it refused to join its smaller company colleagues in acceding, its share of production in the Oasis consortium was promptly nationalized. Shell followed a few weeks later.

Sir David Barran's argument from the start had been that, once the companies gave way to Libya's demands, then the other Middle Eastern countries would be bound to follow suit, forcing a 'leapfrog' in demands between the Mediterranean exporters and those of the Gulf. He was right. When OPEC next met in Caracas in December 1970, it was with both a sense of new exhilaration at a changing world and a determination by the more conservative countries not to be left out. 'Libya took the initiative,' as Gadaffi put it, 'because Libya is a revolutionary country and because we are revolutionaries. Afterwards those who hesitated and those who were afraid followed this Libyan initiative and followed the Libyan storm.'

The Oil Companies Fight Back

When the OPEC conference decided on 'concerted and simultaneous action' to increase tax levels to a minimum 55 per cent and to raise posted prices and eliminate discounts, and when Libya followed this up with a demand for a further 50-cent increase, Sir David knew that this time the companies had to respond. In a 'New Year Letter' addressed to all companies in Libya, he suggested a meeting at which they should adopt a common front and refuse to negotiate with the oil producers except on a combined Gulf–Mediterranean basis. The letter was carefully prepared and successfully brought together twenty-three oil companies on 11 January 1971, in the offices of New York lawyer John Jay McCloy. All the majors were there together with the French, the Belgians, the Germans and the American independents. With American Government officials checking the moves to clear them of anti-trust, the companies decided to stick together. Those who had for so long refused even to recognize the Organization of Petroleum Exporting Countries as a body now insisted that they themselves would not negotiate with anyone else except as a body.

Their tactics were two-pronged: to pressure the members of OPEC into common talks and to ensure that individual companies could never again be picked off in isolation as Occidental had been in Libya. The former objective was pursued through a letter to OPEC delivered before the next organization meeting due in Tehran stating that:

> ... we cannot further negotiate this development of claims by member countries of OPEC on any other basis than one which reaches a settlement simultaneously with all producing governments concerned. It is is therefore our proposal that any all-embracing negotiation should be commenced between representatives of ourselves and OPEC representing all its members' countries, under which an overall and durable settlement would be achieved.

The latter was to be achieved through a secret 'safety net' agreement guaranteeing that the supplies of any one country squeezed by Libya would be made up by the other countries from whatever sources were available at a reasonable cost.

A task force of senior executives from all the companies, the London Policy Group, was then set up in the basement meeting rooms of BP's Britannic House in London to co-ordinate the negotiations. This, in turn, was backed by a group meeting in New York to review policy decisions and to provide any assistance required. Two negotiating teams were assembled, in line with the regional approach already announced by OPEC in Caracas, one, headed by Lord Strathalmond (the son of the first Strathalmond of BP) to negotiate with the Gulf producers; the other, under George Piercy, who had replaced Howard Page at Esso, to handle the negotiations with Libya. Howard Page, much to his delight, was summoned from a ranch in Arizona to join the Gulf negotiating team as an adviser. The American Government sent out a diplomat, Jack Irwin, with a personal message from President Nixon to the rulers of Saudi Arabia, Iran and Kuwait expressing American concern at any cut in oil supplies.

As a grand design to achieve a settlement to end all settlements the whole enterprise could hardly be counted a success. The Gulf states, as Irwin found, were outraged by the oil companies' letter. There was no way that they could sign an agreement that bound them not to react to any terms that might be negotiated in the Mediterranean. In this they were right enough. None of them had any control over the Algerians or Colonel Gadaffi, who was daily making speeches declaring how corrupt they all were. Public opinion would never stand for their tamely allowing Gadaffi to negotiate better terms than their own, particularly in countries like Saudi Arabia and Iraq with a foot in both the Gulf and the Mediterranean camps. Irwin, instead of persuading the Gulf of the need for a single settlement, came back instead convinced of the opposite himself.

The oil companies tried to keep up the appearance of united negotiations, even sending Piercy and his team to deliver their offer under the door of the oil ministry in Tripoli since the Libyans had refused to accept their validity as a negotiating team. But the companies knew that this part of the game was up. All they could do was negotiate as good a deal as they could in Tehran and hope that Libya's follow-up would not make it impossible for them to implement it. Even then agreement in the Gulf was only reached after the initial talks had failed, largely on the question of guarantees against leapfrogging, and the OPEC countries met on their own in Tehran to threaten to impose the terms unilaterally and refuse oil to any customer who wouldn't accept them. Within hours of the 15 February deadline, the companies finally agreed to a five-year settlement that gave the Gulf states an immediate 35-cent increase on posting plus an annual inflation increment, a package that promised to raise their revenues by nearly 40 per cent over the term of the agreement.

Tehran and Tripoli

Libya and Algeria, which had been waiting on the sidelines, then did exactly as the oil companies had feared. They jumped in with even bigger demands of their own. The Tehran Agreement, declared Major Jalloud, didn't go far enough although 'it did represent the first successful effort at unity by the various governments to restore the people's rights, and to that extent the joint stand of the producing countries against the industrialized countries and their monopolistic companies was a victory. . . . The Tehran Agreement will allow us to increase our income considerably, but we shall not be satisfied with what was obtained in that agreement.'[12] A few days later Libyan radical fervour was buttressed by the decision of the Algerians to nationalize 51 per cent of all the French oil interests in their country. Libya was chosen to negotiate on behalf of the four Mediterranean countries – itself, Algeria, Saudi Arabia and Iraq – and started the ball rolling with a demand for a $12 per barrel, or 50 per cent, increase in postings. The companies responded with half that. Negotiations became bogged down on the question of how much any increase was to represent permanent 'differentials' with the Gulf and how far temporary freight advantages. Once again it was the threat of unilateral action that forced the companies to the point of agreement. The short-haul countries gained a 90-cent rise in postings and a 46 per cent increase in government revenue, as well as other concessions, which took them substantially beyond what they would have got if the terms of the Tehran Agreement with the Gulf countries had simply been translated to the Mediterranean.

The 1970 negotiations saw the beginnings of a decisive shift in the balance of power away from the oil companies to the producers. The agreements, of course, never lasted their full five-year term. In some ways it was surprising that they lasted as long as they did – a tribute in part to the skill of Sheikh Yamani, who emerged during the talks as a major influence for moderation in the negotiations both in the Gulf and the Mediterranean. Although the Gulf countries didn't immediately break back to gain equivalent terms to Libya, the talks had shown that the market was moving from surplus to shortage and that, when this happened, the producers would be ready to take full advantage. There were also accidents which strengthened their position: the coincidence of radical regimes in all three major exporting countries in the Mediterranean, the temporary closure of Tapline (it was restored by February 1970), the prolonged closure of the Suez Canal and the sheer pressure of the relentless rise in consumer demand against available supply.

Whatever the reason, however, the demands of the radicals were underpinned by the fact that in the marketplace the 'spot' price of oil (the oil available on the market outside the integrated systems of the major companies) had moved ahead of posting for the first time in a decade. And when that happened, the producers were going to exploit the situation to the full. 'When the Americans came to Saudi Arabia,' Tariki had complained a decade before, 'the King treated them as friends. The idea that a company would always be out for what it could get at the lowest possible price and would treat the government as a natural foe to be exploited whenever possible never occurred to him.'[13] Jim Akins, for a time American Ambassador to Saudi Arabia, vividly remembers the head of one of the American majors waiting desperately for a decision from the Oil Minister, who was sitting inside playing cards with his friends. After about four hours, he was finally shown in to conduct his business, which took all of ninety seconds. 'I spoke to the minister quite harshly,' says Akins, 'at the apparent rudeness and he smiled very sweetly and said, "This is the way they treated us in the past and this is the way we treat them now." ' The boot was on the other foot and the producers were quite willing to use it.

Ironically it was the Saudis who finally broke up the Tehran Agreement and helped set off, to some extent unwittingly, a far greater round of price increases three and a half years later when President Sadat crossed the Suez Canal and started the Yom Kippur War with the Israelis in October 1973. By that time the market had shifted fundamentally. Demand was beginning to outpace the rate of new discoveries and Sir David Barran was publicly warning that, at any rate by the end of the century, the world would be running out of oil if it did not either find more or conserve more. The consumer, he declared in October 1971, was 'peering down the muzzle of a gun'.

The more immediate supply situation was made worse by the sudden drop in American output for the first time in a century which had been exacerbated by a five-year delay in getting the Alaskan pipeline project through the planning process. A number of producers, including Kuwait and Libya, had now cut back production in order to lengthen the life of their reserves and the world's additional requirements for oil, including those of the United States, were now concentrated almost exclusively on two countries: Saudi Arabia and Iran. The basis of the Tehran and Tripoli agreements had, in the meantime, been undermined by the fall in the dollar on the world's exchanges and the acceleration in the rate of inflation which had greatly reduced the purchasing power of the producers' revenues. To add to the confusion in the market, the producers had negotiated a series of deals throughout the Gulf giving them participation in the concessions and thus oil to sell on their own account.

Participation was very much the particular dream of Sheikh Yamani, presented to the companies both as an alternative to the outright nationalization then taking place in Libya, Algeria and Iraq and as a means, in Yamani's words, of creating an indissoluble bond between governments and companies 'like a Catholic marriage'. The technology and skills of the oil companies were now to be harnessed to the needs of the oil producers through agreements which gave the state an initial 25 per cent of the equity in Aramco and other consortia, rising over a few years to majority control. This was a gradualist rather than a revolutionary approach. What it did in a more technical sense, on the other hand, was to bring to market new sources of crude oil just as demand was taking off. To their astonishment, and irritation, the producers found that their first sales of oil fetched prices far higher than they had expected or the companies were paying in taxes and royalties under the old concession arrangements. The companies argued that this was only because they were the first of the state oil sales and therefore attracted high 'entry bids' by purchasers anxious to open arrangements with the key producers. The producers, for their part, suspected that the oil companies had been deceiving them and that the market warranted much higher posting yet again.

The 1973 Yom Kippur War

OPEC had already met and declared its intention of seeking a substantial increase in postings, as much as double, it warned the companies, and the representatives of both sides were just flying into Vienna for the first round of negotiations on 8 October 1973 when Sadat sent his troops across the Suez Canal. The oil companies were cowed. The level of increase being demanded – a rise in posting from around $3 to $6, or a minimum of $5 as Yamani was asking for – was far greater than they felt they could afford to concede. Yet they had gained no support from their respective consumer governments either to resist or concede. The producers, on the other hand, were quite determined to obtain what the market could bear and all the more ebullient because of the apparent first success of the Egyptians in gaining a foothold on the other side of the canal.

The meeting broke up with the oil company representatives, led by George Piercy, declaring, against the pleas of Sheikh Yamani to continue discussions, that they had to consult their governments formally before they could possibly increase their offer. The producers left to consider their action and Sheikh Yamani returned to Riyadh to prepare the first moves in a separate but parallel issue, the use of oil as a weapon to support the Arab cause over Palestine.

Just how far the two issues of price and Palestine were connected in the Saudi and Arab mind at the time has been the subject of endless argument ever since. Sheikh Yamani has always denied it, and, in their origins at least, the two probably did start off as distinct issues. The question of Israel's occupation of Arab territory in the 1967 war was always of an order of importance in King Feisal of Saudi Arabia's mind far and away above any question of oil policy or price. As the Keeper of the Holy Places of Mecca and Medina he felt bitter about the Israeli conquest of Old Jerusalem and the spot where Mohammed had ascended to heaven. The first efforts by the Arabs to embargo oil going to the 'friends of Israel' after the Six-Day War had ended in dismal failure. Feisal simply pursued the issue with greater diplomatic determination. When words failed to move the United States and her allies, he returned to the oil weapon. All through 1973 he sent fruitless messages via the American Ambassador, via Aramco and via Sheikh Yamani to President Nixon, warning him that he would have to use his oil muscle if there was no move by the United States. In the summer of 1973, on a visit to Switzerland, he even summoned the heads of the four Aramco partners – Exxon, Texaco, Socal and Mobil – to repeat his concern. There was no response. The State Department simply took the view that they had heard it all before and that anyway it was just empty rhetoric. In August President Sadat secretly flew to Riyadh to seek Feisal's support for his intended attack on Israel. Feisal promised him money and the unsheathing of the oil weapon. Yamani was asked to work out details of how effective pressure could be brought to bear on the United States to force it to exercise its influence on Israel by at least restraining the increases in Saudi production being demanded if not by starting to reduce supplies altogether.

However distinct the Palestine and oil price questions may have been, President Sadat's October war brought them straight together. As the first military moves seemed to go in favour of the Arabs, President Nixon rushed in military aid to the Israelis. King Feisal once again appealed to the President to remain neutral in the conflict by sending his Foreign Minister and three other Arab ministers to Washington in mid-October. They came away empty handed, to an Arab public opinion made all the more resentful by the reversal in military fortunes in favour of the Israelis.

On 16 October the six Gulf Oil Ministers met in Kuwait to announce their unilateral intention to raise the posted price of oil by 70 per cent to $5·119 and to keep it 40 per cent above the level of open-market prices for the future. The next day the Oil Ministers of the ten countries belonging to the Organization of Arab Petroleum Exporting Countries (OAPEC), founded in 1968, met in the same place to announce an embargo on oil shipments to all those considered to be favourable to Israel. To give the embargo teeth, unlike the previous occasion, oil supplies were to be progressively reduced by 5 per cent per month, a figure raised at the next meeting to an immediate 25 per cent cut. Within that total, favoured countries were to have all the oil they wanted and preferred countries (those which had broken off diplomatic relations with Israel) were to get the same as they had been before the embargo as were a few countries considered sypmathetic to the Arab cause such as Britain and France. Other countries were to be reduced in line with the general reduction or, like the United States and Holland, cut off completely. The bite was there.

The effect on the industry and the oil market was traumatic. Aramco itself had little choice but to obey the orders, at least so far as volumes were concerned. While some countries, such as the United Arab Emirates, might interpret the OAPEC decisions somewhat liberally, Kuwait and particularly Saudi Arabia did not. All the extra oil that could be squeezed out of non-OAPEC sources such as Iran couldn't make up the difference of a reduction that by December amounted to

4·5 million barrels a day. What is more, the oil companies and the market had to work on the assumption that the progressive contraction of supplies would continue.

The consumer governments were left helpless and divided. They had been caught, apparently unawares, by the sudden price explosion in oil during the year, unwilling to sanction the major price rise negotiated by the companies but unable to think of any alternative. In the immediate aftermath of the crisis, US Secretary of State Henry Kissinger attempted to rally the Western consumers in a counter-OPEC action. However, while agreeing to form the International Energy Agency (IEA) under the aegis of the Organization of Economic Co-operation and Development (OECD), the Europeans refused to make it into a weapon of confrontation with the producers. Indeed, they then proceeded to squabble among themselves over supplies, those like France and Britain, regarded favourably by OAPEC attempting to achieve the full advantage over those like the Dutch who were embargoed. It was left to the companies to attempt to ration out their supplies as best they could between their customers, exchanging where possible non-embargoed oil for embargoed oil to keep supplies evenly reduced – over-reduced as it turned out. In a heated meeting the British Prime Minister, Edward Heath, was told by Sir Eric Drake, chairman of BP, and Sir Frank McFadzean, chairman of Shell, that they would not treat Britain as a special case, whatever the assurances made by the Saudi Arabians.

The spot market, in the meantime, went haywire as purchasers bid up the price of a barrel of any available oil from $10 to $15 and finally a peak of $22. Iran, against strenuous opposition from Saudi Arabia, led the call for yet further OPEC price rises to match and to keep the posted price at the 140 per cent of market price agreed in Tehran. While Yamani tried to hold the posted price back to $8 per barrel, which would give the producer governments a revenue of $5, the Shah insisted on an $8 revenue, requiring a posted price of $13·30. After a tense and impassioned meeting of the OPEC Gulf ministers, which took place in Tehran on 23 December, the producers decided on a rise to $11·651, an increase of nearly 400 per cent in the posted price of oil within three months and a rise from $1·80 to $7 per barrel in their average income. This income level was pushed up yet again to over $10 per barrel by the end of the year as the terms of participation were adjusted to raise the revenue on the remaining oil held by the companies, whose profits were enormously and conspicuously boosted by the price rises.

The End of an Era

The quadrupling of oil prices ended an era in oil. It ended the age of relentless demand growth which had gone on for almost the whole

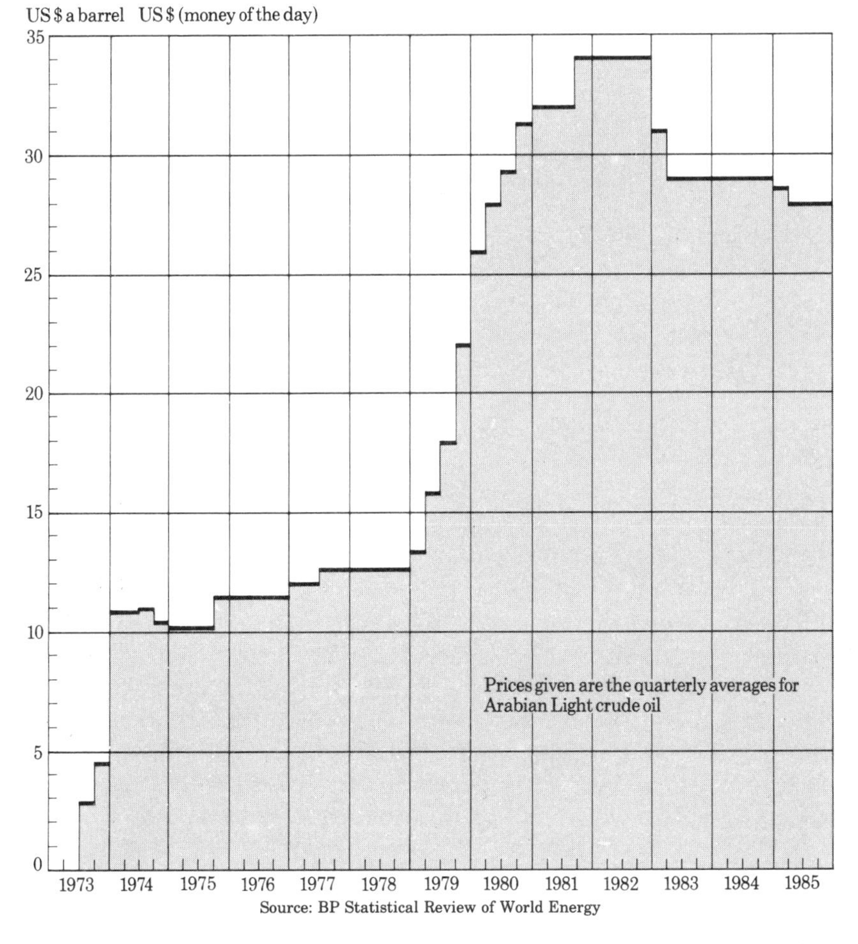

The oil price explosion, 1973–85. The average price of Saudi oil, which stood at $2·80 before the Arab-Israeli war of 1973, had nearly quadrupled by the spring of 1974, before nearly tripling again during the Iranian crisis.

century and ushered in a period of conservation, investment alternatives and profound debate as to whether the industrialized world had somehow reached the limits of its ability to go on using up the natural resources of the earth as it had done ever since the Industrial Revolution. It brought to a close a generation of ever-increasing growth in supplies from the Middle East and opened a new chapter in which the producers had to decide how best to exercise their power, in the political, economic and, within their own boundaries, the social field. It proved both the apogee of the ability of the major oil companies to organize supply and demand over and above national will and controls but also the beginning of the decline in that power. Never again would the oil companies play such a role. Participation and nationalization were depriving them of their supply and forcing them to decide whether they were to act as the agents of the producers or the consumers.

Yet, if the winter of 1973–4 ushered in great dreams and debates as to the transfer of wealth between rich and poor, between raw-material producer and manufacturing user, between the old world of Western industry and the new world of the developing, resource-rich countries, it also set in motion the moves to its own adjustment. Investment in alternative forms of energy more than tripled in the following years. Exploration was undertaken with renewed vigour both in old oil provinces such as Texas and in new oil ones like the North Sea. Consumption fell for two years running in 1974 and 1975. The world was hit by recession combined with inflation which affected the producers as much as their major customers.

The politics of OPEC became increasingly centred on the rivalry between Saudi Arabia and Iran as the two major suppliers of additional oil to the world. Saudi Arabia, in some ways because it had always seen oil not simply as a source of revenue but in wider political terms, and because it had huge reserves, argued that the price should now be stabilized and that opportunism on market conditions should give way to planned price adjustments according to inflation and currency indices. Nor was the kingdom afraid to use the threat of swamping the world with oil to pull its more radical OPEC colleagues back from new price rises. Iran, on the other hand, determined to make the leap into becoming one of the world's top ten industrial nations and pushing itself to the limits of its productive capacity, led the more radical countries in seeking to maximize revenues and push for every little price increase it could obtain. Higher prices, argued the Shah somewhat loftily, would do the consumers good and would teach them to conserve what was 'really a noble fuel' that should be reserved for premium uses only in transport and petrochemicals.

Up, Up and Away

The Shah got his way, although not quite in the manner in which he intended. His decline and fall in late 1978 and in early 1979 occurred just as consumer demand for oil was on the increase again, as the first impact of higher oil prices was eroded by inflation, as a number of companies had announced their intention to cut output for 'conservation' reasons and as supplies for the winter appeared to be getting tight. The fall-off and then complete stoppage of Iranian oil exports had the markets once again bidding up the price, this time to $40 per barrel and more. The oil companies and consumer governments, remembering the petrol queues of 1973, grossly overreacted. Oil companies rushed to build up stocks in expectation of a shortage when the volume of government-held stocks was more than adequate to meet any emergency. Governments refused to release stocks in a move that might have controlled prices because of their fears for the future. Unfortunately the shortages remained just below the levels that would have triggered the automatic IEA sharing agreements.

OPEC had already raised prices to $28 per barrel in June when the spot market for oil started to move up rapidly. When it met again in Caracas in December 1979, strenuous efforts by Saudi Arabia to find a new price, around which stability could be restored, failed. The Saudis, in co-operation with the Venezuelans, moved their prices up $6 a barrel, only to find the radicals moving up still further. Acting in almost total disarray, individual countries sought to raise prices simply until they seemed to stick and it was not until Iran tried for $35 in April 1980 and failed to make it hold that levels finally came back into some kind of alignment at $34 per barrel, thus doubling the 1974 prices again.

In his fall, the Shah of Iran had ironically achieved the fullest expression of his oil policy. As had other high-population countries, he had persistently sought to maximize his revenues by pushing the price to the highest possible level. Western critics accused him of greed and carpet bargaining ethics. With a country of 40 million people, however, perhaps as many as a third of them nomads, dependent in the past on the thin earnings of agriculture and crafts, oil seemed the only means to growth. And the higher the price, the bigger the income. OPEC's other large-population countries – Algeria with 20 million; Nigeria with 82 million; Venezuela with 13·5 million; Iraq with 12 million; and Indonesia, with a staggering 140 million – all adopted a similar line at OPEC meetings. The problem, as the Shah found out to his cost, was that development cannot be accelerated at too rapid a rate without imposing intolerable strains on the society. Land prices soared in Tehran, making a few very rich and the poor homeless. Sucked into the cities, the nomads had no work waiting for them. The ports clogged up with imports. The latest computers lay idle for lack of technicians. The country was encouraged to spend on more sophisticated armaments than most countries in Europe. The Shi'ite revolution of Ayatollah Khomeini terrified most of its neighbours, not simply because so many of them had Shi'ite minorities of their own (Saudi Arabia itself experienced a shaky moment no long after when religious extremists took over the Holy Mosque of Mecca), but also because the return to Islam and its virulent rejection of all things Western held, and still holds, a powerful moral sway throughout the Middle East, among all sects.

For Saudi Arabia, on the other hand, and its Oil Minister, Sheikh Yamani, 1979–80 represented a profound reversal of oil policy. Just how far Yamani has been the architect and how far the messenger of Saudi policy within OPEC has been the subject of constant analysis by outsiders. Sir David Barran describes him as, '. . . one of the really great men of this century . . . a very, very remarkable man with a real grasp of economics, a real grasp of world requirements, with the great saving grace of an enormously good sense of humour.' Dr Parra says he is 'a remarkable man. He has been able to keep together forces which have worked in opposite directions. . . . He has had to face, over the span of twenty-five years, enormously different situations. He has worked within his own country with a number of succeeding administrations and heads of state. He has had to convince his own constituency of his views and policies and he has had to face within the organization a diverse group of countries, countries with minimal reserves, countries that wish to have the highest prices under any conditions possible, countries which are unfriendly to Saudi Arabia and others which form part of the Gulf group of countries. . . . I believe that Yamani is a force for stabilization. I think he is essentially an optimist and that he looks at the outlook not as a photograph but with a broader and longer term view of things.'

Others, like the *Wall Street Journal*'s chief oil reporter, Youssef Ibrahim, see him more as the diplomatic mouthpiece of the Saudi royal family. 'Yamani is in an astonishing position,' concedes the author, Anthony Sampson, 'being a commoner in a country of princes with a very uncertain relationship with the King. At the time when in fact he decided to hold together with OPEC in 1973 and allow the price to be quadrupled, he did then try to get through to the King and could not do so. And the King later made it clear that he took the opposite view, that Yamani should've stayed out of OPEC and kept the price down. Nevertheless Yamani was so influential, so intelligent and so diplomatically adept, particularly at keeping the Americans happy as well as keeping the price up, that his position has remained fairly secure in spite of all rumours to the contrary.'

Whatever the differences between himself and the royal family – and they have been sharp at times – Yamani has consistently and accurately given voice to the view of the moderates within OPEC, the high-reserve, low-population countries of the Gulf such as Kuwait, the United Arab Emirates, and Saudi Arabia. For them, the opportunities to diversify away from oil are limited and the need for a stable market environment over the long term is all the more important. Even before the 1973 price hike, the Gulf countries had become rich, their purchases of English country estates and Park Avenue apartments legendary. In the wake of 1973, they have tried hard to diversify their sources of income. The Kuwaiti Government has become a major investor in oil marketing and refining facilities in Europe and the United States as well as investing directly in foreign companies. Saudi Arabia has built up a huge petrochemical and refining industry to add value to oil for the future and has attempted to use its low raw-material costs as well to build up an agricultural business based on fertilizers. Wheat and vegetables are literally grown on fertilizer watered by desalination plants.

For the Saudis, as well as the world at large, the 1979–80 round of oil price increases was the shock that should never have happened. It forced the consuming nations into reducing consumption and finding alternatives to Middle East oil. It induced a profound hostility to the Arab cause in American public opinion and it depressed the markets necessary to sell the products of Saudi Arabia's grandiose industrialization schemes. Suddenly the consumers no longer wanted the oil-related products in which the Saudis had so heavily invested. 'It was a shock,' says Prince Abdullah Bin Feisal Bin Turki, who heads the country's petrochemical endeavours, 'to realize that after ten years of building infrastructure and factories – and after almost all countries in the world participated in the building of this because we brought along a lot of people from outside, we bought a lot of equipment and materials . . . the demand had slowed down and we had to slow our plans.'

It was a shock big enough to induce Saudi Arabia to abandon its policy of supporting OPEC price structure. As demand fell during the early 1980s and as prices in the market weakened, Sheikh Yamani led the effort to try and hold OPEC together by reducing output and refusing to sell at a discount. The implication was a continuously reduced production level. Saudi Arabia, which had raised production to prevent prices rising too much in the 1974–8 period and then again in 1980, now found itself taking the greatest burden of contraction. By 1985, when its output had collapsed from a peak of 11 million barrels per day to barely more than 2 million – less than half the level allowed under the OPEC production regulation agreements – the Royal Council stepped in to say it had had enough. Sheikh Yamani was instructed to prepare contracts to sell oil at whatever price it would fetch in the market. Over the winter of 1985–6, Saudi output doubled and prices fell through the floor. Saudi Arabia was prepared to support a new effort at price stability, Sheikh Yamani declared to his OPEC colleagues and to the outside world, but only if everybody joined in, including the non-OPEC producers. 'Now the consumers,' he argues, 'are committing the same mistakes OPEC did by raising the price of oil. You are pushing the price of oil as low as it is possible which will bring life to OPEC and make it a very strong organization later on.'

5
The Independents

Just after Easter 1986, with oil prices plunging below $10 per barrel almost to their pre-1973 crisis levels, the American Vice-President, George Bush, set off on a tour of Saudi Arabia and other countries in the Gulf. To the astonishment of many of his colleagues and even the President, who had been busily arguing what a good thing the fall in oil prices was, Bush announced that he would be telling the Saudi Arabians that the fall had gone far enough. It was now jeopardizing American security by battering the nation's oil producers.

'It is important that we have a strong domestic industry,' declared Bush. 'We are reaching a point,' stated Energy Secretary John Herrington, 'where we need to be concerned. Service companies, drillers, producers are all having troubles. Banks that lend the money – terrible problems.'

'I think it is essential,' argued Bush, in words that came perilously close to the sentiments of Sheikh Yamani, whom he was going to see, 'that we talk about stability and that we do not just have a continued free fall like a parachutist jumping out without a parachute.'[1]

The reason for the sudden signs of policy reversal among the ranks of the US Administration – quickly denied and smoothed over by President Reagan himself – was not hard to find. Back home in Texas and the southwest, where oil was business and the business was oil, the price collapse was causing 'devastation' in Bush's words. And he ought to have known for it was in Texas that Bush, the Ivy League son of a northeast stockbroker and senator, had made his first, and his second, million by building up an oil drilling and exploration company that was now threatened with collapse if the fall in prices continued. Bush had long sold out in pursuit of a political career but his son was still in the exploration game and Texas remained his political base, as it was for his friend and supporter, James Baker, former White House Chief of Staff and now US Treasury Secretary.

Up in the northeast and the Midwest, where the oil price rises and the accompanying decline of the old smokestack industries had thrown several million out of work, oil simply represented unavoidably high prices. When the oil crisis hit the United States in the winters of 1973 and 1979, the oil-importing states had cried out for government aid in keeping prices down and enforcing conservation. The Texans had gone around with car-bumper stickers declaring, 'Let the Bastards Freeze in the Dark' and 'Drive at Ninety and Freeze a Yankee'. Now the tables were turned. It was Texas that was appealing for government aid, as its Governor publicly called for an oil-import fee to protect the home industry, the number of drillers working in the United States collapsed from a peak of nearly 4000 in 1981 to barely more than 1000 in March of 1986 and as the banking authorities, struggling with some 400 southwestern banks pushed on to their 'problem bank' list, tried to prevent a stream of failures because of the added energy problem. 'It's the Frost Belt's turn to gloat,' wrote Michael Kinsley, one of the leading American right-wing commentators, on the *Washington Post*. 'Let 'em rot in the sun.'

The Hunts and their Texas Millions

Oil is Texas to Americans and to the 50 million-odd viewers of the *Dallas* television soap opera: the symbol of riches, greed and the perpetual dream of wealth spouting from the desert where that extra foot of drilling makes the difference between becoming a millionaire or a pauper. And wealth is the dream of the drillers and riggers who fly in and out of the oil towns of the world from Balikpapan in Borneo to Aberdeen in Scotland. It is no fantasy. At the same time as the fall of the Shah set off a major crisis in the oil markets in late 1979 and early 1980, the world was treated to an even more dramatic rise and then almost total collapse in the price of silver. The rise had apparently been engineered by the Hunt brothers from Dallas, one of the richest oil families in Texas, if not *the* richest. Their failure required a $1 billion loan to bail them out, the largest loan ever organized to save individuals. Hauled before a congressional committee, two of the Hunt brothers, Nelson Bunker Hunt and his younger brother, William Herbert, who had been partners with BP in Libya until the concession was nationalized, blandly insisted that they had no precise idea of what their silver holdings were or how much they had put into hoarding the precious metal. 'It is incredible,' retorted the chairman of the committee, Congressman Ben Rosenthal, 'how neither of you have any notion of how much you are worth, or the Hunts are worth, or anybody.'

'No, I don't have any idea,' answered Bunker Hunt casually. 'A fellow asked me that once and I said, "I don't know, but I do know people who know how much they are worth generally aren't worth very much." '[2]

It was not such a jocular reply, although Hunt gained enormous publicity from it and frequently repeated it. The Hunts probably did not know how much they were worth, at least not until they were forced to mortgage everything from Nelson Bunker Hunt's string of racehorses to his brother Lamar's collection of Napoleonic relics to back the loan. Their father, Haroldsen Lafayette Hunt, or H.L., as he was always known, was called the 'richest man in the United States' by *Life* magazine in 1948, a fact which he denied despite a personal fortune worth, on his own admission a few years later, some $2 billion and an annual income, which he wouldn't reveal, of over $50 million.[3]

H. L. Hunt was indeed a character straight out of *Dallas*, a bigamist with not just two but *three* families at the same time. He had made his fortune out of buying, under circumstances that were not quite above board, an East Texan oilfield, perhaps the most lucrative field ever discovered in the United States. He had started life as a gambler in the logging camps of Arkansas and on the riverboats of the Mississippi; professional cards require patience, a sense of your opponents' weakness and a phenomenal memory, all of which Hunt possessed in plenty.

His son by his third marriage, Ray Hunt, recalls being deeply impressed when on one occasion he was at a reception and someone came up to his father and asked him what it felt like to be known as 'the world's greatest gambler'. 'You've taken all these great risks, you've got a reputation of being a fantastic gambler and very successful, how does that feel?' H.L. dismissed the man, saying the reputation was completely false.

His son was puzzled and, turning to him as they left the reception, he asked, 'Dad, how in the world can you possibly have answered that man saying that you couldn't consider yourself to be a gambler, with all the different things that I know you've done?'

'And he said, "Ray, I have never taken a risk, but that if everything went against me I couldn't keep right on going."

'Now I'm sure he forgot some of the gambles and risks he took as a very young man. But that made a very big impression on me and I think that really is the secret to success. As an independent in the oil and gas business, you don't take risks where there is so many eggs in one basket. And if everything goes against it, you keep right on going.'

H.L. himself, in one of the many pieces he wrote in his later years put it this way: 'Most business transactions are a gamble, some good and in others the odds are bad. The percentage of success must be calculated and the deal made only if the reward will justify the danger of the loss. No deal should be made if the occurrence of a loss would be catastrophic, for a better time will come.'[4]

They were words that his sons might have done better to heed when they embarked on their great silver-buying spree in the 1970s. It was an approach admirably suited to oil drilling, which H.L. took up with varying degrees of success in Arkansas in the 1920s before moving to Texas as the Depression began to sweep across the United States. There he subsequently turned up in Rusk County, East Texas just in time to witness the discovery of the mammoth East Texas oilfield by Columbus M. Joiner.

'Dad' Joiner and the Daisy Bradford Well

'Dad' Joiner, as he became almost universally known after the discovery, was the archetypal wildcatter of old America, a man then aged 70 who had been drilling around the southwest for some twenty years, finding oil here, drilling dry holes there, buying leases and selling them, failing by a few hundred feet and for want of $1000 to discover the Seminole Field in Oklahoma. None the less he was ever friendly and imbued with an overwhelming sense of optimism.

A newspaper editor in John Ford's classic film *The Man Who Shot Liberty Valance*, on hearing the true story of how the main character became a hero, declares: 'This is the West. If it's a legend, we print the legend.' The legend of the discovery of the Daisy Bradford well, as related by Dad Joiner, was that one day, while in the Texas port of Galveston on the Gulf of Mexico, he had dreamed that he was destined to discover the biggest oilfield in the world. Why this should have taken him several hundred miles inland is not clear, but afterwards he liked to say that it was in search of the hilly countryside with a creek that he had seen in his vision. He claimed that he had finally espied it at a farm owned by a widow, Daisy Bradford, and that he had written to her explaining his dream, promising that if he could drill on her land he would find a field that would allow her to stop working for ever. The truth is probably a little more mundane. He had taken a large number of leases in the East Texas area because they were cheap and a geologist friend had expressed the hope that there might be an extension of a known oil basin there.

Joiner chose to start drilling on the Daisy Bradford property only because she had sold him the lease on condition that he began there first. Forced to use makeshift equipment and local farm labour, the first two wells both had to be abandoned when the equipment broke and the drill bit got stuck down the hole. Joiner's funds had been exhausted several times and he was only able to meet the wage bill and renew the concessions by hawking 1 and ½ per cent shares around the local community. At one time, on the recollection of Jim Maxwell, one of his workers, he even had to go as far afield as California to sell the shares. The wages had to be brought up by the bank manager personally, according to Maxwell, since Joiner could not write a cheque. In the Depression, however, jobs were hard to come by and the wellhands and local investors could not but look on the enterprise with the same kind of passionate optimism as Joiner. Finally a third well was started at a spot near the second for no other reason than that the land was flat and Joiner didn't have the money to clear space on a slope. He was lucky. Had he drilled a few hundred yards to the east, where he originally wanted to, the well might easily have been dry as others were after it.

As it was, drilling the third time proved easier as well as lucky, although it still took thirteen months to get to the depth he wanted. At almost 1066 metres (3500 feet) Joiner took a sample of rock and found it to be saturated with oil. Aware that some scouts from an oil company were about, he apparently left the sample lying around on the assumption – correct as it turned out – that they would therefore never believe that it was the real thing. Word had leaked out though and one of those to hear of it was H. L. Hunt, who had just moved to Dallas having managed to make some money out of an oil find in Arkansas. Tipped off, Hunt rushed 150 miles to the site where he tried to get in on the action, if only succeeding in getting into the famous celebratory photograph.

Hunt was not the only one excited by the find. The day that he arrived in September 1930, Joiner came across the first signs of oil in his drilling. When news of this broke, the area went wild. People, many of them with shares or thinking they had shares in the discovery, came from hundreds of miles around to watch the final progress of the well. Shares were traded dozens of times over. Every available piece of land in the vicinity was snapped up, not least by Hunt himself. Joiner very nearly became the victim of a court order to wind his operation up brought by investors who were outraged by the growing evidence that he had sold shares in his lease several times over. His allies soon spread the word that it was a stratagem of slick city lawyers hired by the majors anxious to get their hands on the property, a surefire appeal in a Texas still proud of its independent background. Work was renewed on the drilling until, on the evening of 5 October, a gurgling sound was heard from the well. 'There was about seven or eight hundred people there,' Maxwell remembers, 'and the cars was lined up about a mile and a half along the side of the road in a cotton field. Everybody just felt chills running down their back and everybody was real happy, hollering when the oil went over the derrick. It just gushed out . . . and everybody screamed and that old black oil just come down on the derrick and it just made you feel plum funny.'

As had happened so often before in the American oil business, with the discovery came the real problems, mostly in the form of money and legal claims. One well didn't make an oilfield and within a few days of the first gusher, the pressure in the well dropped sharply. Other wells drilled nearby to the east proved to be dry. It was only Texas sentiment, and a judge who declared, 'I believe when it takes a man three and a half years to find a baby, he ought to be able to rock it for a while', that got a receivership action postponed indefinitely. Hunt, who was now convinced that the Daisy Bradford well was on the eastern edge of a potentially major field (hence its odd behaviour and the dry holes to the east), moved in with an offer to buy out Joiner completely. Hunt was later claimed to have suborned the drilling supervisor on a critical well being drilled by another company to the west to allow one of his scouts to check its progress. Finding Joiner hiding from his angry investors in Dallas, Hunt negotiated with him through two long, all-night sessions until he finally emerged with an agreement to buy the old man out for $1·335 million with a down payment of $30,000 in cash, the rest to come out of royalties. During the talks Hunt received news from the scout that the other well was showing signs of oil.

Joiner subsequently accused Hunt of keeping that information from him. He also took action against him for 'fraud'. But that was over two years later. For the moment he appeared happy enough, indeed he seemed positively to have liked and admired Hunt, sensing in him a stronger and more youthful man after his own spirit. The deal got the creditors off his back, absolved him of legal claims and gave him a sizable amount of cash to spend doing the town with his secretary. Hunt, penniless himself at the time, had still to raise the money, which his own bank would not lend him, to drill on his newly acquired leases, as well as having to fend off the 200-odd claimants to the rights, any one of whom could have tied him up for years. He bought them off, as he had Joiner, with cash. He may well have paid Joiner off a second time for, when Joiner, egged on by others, filed suit against him at the end of 1932, the two went into another prolonged huddle. As the court hearing started, Joiner got up to read a prepared statement which said that it was all a mistake and he had never been deceived or defrauded. The bemused judge then closed the case and Joiner never said a word to explain his change of heart, going off to search for new giant discoveries in North and West Texas and finally returning to Dallas, where he died in 1947 aged 87.

'You know, that's one thing I honestly believe,' says Sandy Joiner Roberts, Dad's great-granddaughter, who runs a restaurant patronized by some of the Hunts. 'I believe that they would have always kept their word with one another. Whatever happened in the hotel . . . I believe each was trying to act in an honourable way, the handshake. I think they understood one another in that respect and there's no doubt in my mind that there was a good reason for Hunt to have ended up with the money.

'I'm not so sure he was "poor old Dad Joiner", you know. In his own way, I believe he was happy with himself. From letters I've read and things I've heard in the family, he was doing what he wanted to do. He was still seeking something that was out there. He really couldn't get his hands on it and I think maybe that was his destiny. . . . He was in his mid-eighties when he died and he was still promoting and looking for another well like Daisy Bradford. There must have been something deeper than deep about the gambling spirit that was in him.'

The thought that her family might have been as rich as the Hunts doesn't worry her. What does is 'the old famous picture of Dad Joiner,

Doc Lloyd and Daisy Bradford. If Hunt is in the picture especially. You know, I keep thinking, move over; let the man who brought this in have his day because he's not really asking for that much. You know, fate has written this thing out that it won't last that long. So give the man a front-page story. And I want Hunt out of the picture sometimes. In that respect I think of him as being more of an opportunist getting in that picture.'

The Richest Man in the United States

Opportunist or not, H. L. Hunt never looked back. By the time he acquired Joiner's leases, he had already got into the pipeline business in the area. When Joiner took him to court, enough drilling had already been done to establish the East Texas field as the largest in the world with recoverable reserves that have by now produced 4 billion barrels and are still pumping it out. Joiner's leases only amounted to around 4 per cent of the field. This was still enough to earn Hunt $½ million in his first two years, despite the collapse in oil prices from over $1 to 10 cents (partly brought about by the East Texas oil discovery itself) and despite the limited success of attempts, encouraged by Hunt, to restrict production in an effort to sustain prices. Hunt drilled and overdrilled with the best of them and probably flouted the controls with 'hot oil', smuggled out like bootlegged liquor into the night.

Over the next two decades Hunt greatly expanded his oil empire, starting an oil-drilling company, taking over a refinery and buying new concessions throughout the southwest and California. He remained an inveterate gambler, putting so much on Saturday football games that teams of employees had to spread the bets among the bookies, none of whom could cope with the entire sum. *Life*'s decision to pick him out as potentially the richest man in the United States in 1948 found him, to the surprise of his friends and family, ready to emerge from his secretive corner. Suddenly he became a public figure. And he enjoyed it. Eccentric enough in some of his personal obsessions – he was a health food fanatic before his time, milling and making up his own food and going to work each day with a nut cutlet wrapped in a brown paper bag – he now fancied himself as a sort of home-spun philosopher preaching the virtues of American capitalism. He financed a programme called *Lifeline*, broadcast on more than 550 stations around the United States, and wrote newspaper columns, inveighing against the creeping threat of Communism, supporting Senator Joe McCarthy and attacking American foreign policy as too soft. When President Kennedy was assassinated in Dallas in 1963, the Hunts were immediately picked out by the press as being implicated, partly because Nelson Bunker Hunt had helped pay for a full-page advertisement that morning welcoming the President in stridently sarcastic and critical terms.

In fact, as his widow, Mrs Ruth Ray Hunt, retorts, the accusation was unfair. 'My husband had voted for Kennedy, he knew his father, he was a self-made man and felt that Kennedy jun. would certainly make a fine president.' Although Hunt liked fame – 'I'm a notoriety seeker now,' he proclaimed when the newspaper attention started – he was never that enamoured with wealth as such. 'Money is nothing. It's only a way of keeping score,' he used to say. He took the attitude of many a man who has lost money as well as won it: that it deserved respect and needed to be controlled. Once, according to his widow, he 'gathered the children into the library and wanted each one to understand the cost of electricity, per wattage, how much it cost to leave that light burning for hours'. The children had difficulty in restraining their laughter. He hated spending money. When he finally went on a tour of Europe, he refused to shop, telling his wife that what he enjoyed most was the feeling of walking into a place and, as she recalls, 'seeing how much he could walk out without, how much he could leave behind'. He also hated wasting time. At the height of his success someone worked out how much he was earning per hour. He felt the sum could be greatly increased if he could have a working lunch over his nut cutlet.

All this time 'Popsy', as he became known in his later years, was keeping no less than three families going. He kept his first wife, schoolteacher, Lyda Bunker Hunt, in El Dorado, Arkansas for a long time before moving her to Texas. He had seven children by her between 1915 and 1932, including four sons: Haroldsen Lafayette III, known as Hassie, Nelson Bunker, William Herbert and Lamar. 'Hassie', his favourite, turned out to have his father's nose for oil, making several million on his own account, but not his stability. At the age of 21, after years of increasing bouts of violence and tantrums, he was lobotomized, in the manner of the time, to live on as a kind of ghost. Nelson Bunker and William Herbert went into the oil business with Placid Oil. Lamar, the youngest, became a sports promoter, making a name for himself by investing in his own American football team, the Kansas City Chiefs, a soccer team and basketball as well as founding the World Championship Tennis Circus which revolutionized the sport.

In 1925, while hustling for real-estate deals in Florida, H. L. Hunt met a young lady, Frania Tye, in Tampa; conducted a whirlwind romance and was married by, so he claimed, a judge in his home, under the name of Major Franklin Hunt. When Frania became pregnant with their first child, he set her up in a home in Shreveport, at the time of the Joiner deal. By 1934 she had had four children by him, two girls and two boys. He visited fairly regularly for several days to a week at a time. On her account, she had no idea of Hunt's bigamy until told by a friend in 1934. Even that did not stop the relationship. Hunt moved

Frania first to New York and continued this second married life, even trying to convince her to become a Mormon to regularize the affair, and then back to Texas so she could be closer to him in Dallas. It couldn't last without a scandal, especially since the two wives were now only 250 miles apart. In a rage one day Frania brought the children over to Dallas and dumped them on him. Hunt realized that the time had come to sort things out so he brought his two wives together. They met separately, Frania then departing for Los Angeles where Hunt paid for his children's education. In 1942 Frania came to Dallas with her lawyer to arrange a more formal settlement. The talks took place in a hotel and, as with Joiner, Hunt took pains to conduct some of them without anyone else present. The result was a settlement of $300,000 in cash and a monthly cheque. Within a fortnight Frania was married to a Colonel John Lee, some said as part of the settlement.

At the same time, – coincidentally or not – Hunt started dating a secretary from his Shreveport offices, Ruth Eileen Ray, a 25-year-old country girl from Oklahoma. She too became pregnant and was shipped off to an expensive apartment in New York. After having his son she moved to Dallas as 'Mrs Ruth E. Wright', living within walking distance of his grand porticoed southern-style mansion called Mount Vernon after George Washington's home (it was somewhat larger). There she had three more children, all daughters, giving Hunt a total progeny of fifteen, one of whom had died in childbirth.

'Hunt,' explains the author Harry Hurt whose researches have uncovered the extraordinary story of the family in detail, 'thought he had this genius gene. He thought that by propagating his own species and by having children, he was doing the rest of the world a favour. Passing on this genius gene. His belief developed from when he was a young man. He was very good at card tricks and could do mathematical computations in his head with great speed. Later he became convinced that he had an oil-finding sense. He first called himself a "creek-ologist", meaning that he would look at the way the rivers and creeks in a certain area would flow and use that as a means of deciding where to drill for oil. Of course, a lot of the modern science of geology discounts the topography of an area as giving any indication of what's underneath. But Hunt nevertheless went on this theory and later his sons showed perhaps they had got, or inherited, some of this themselves.'

Hunt may or may not have believed the genius gene theory. He was cranky enough in his way. But he also had a dry sense of humour and a twinkle in his eye. Women seemed to find him genuinely fun to be with and he with them. Ruth Ray was more than a match for him. When his first wife, Lyda, died, Ruth and her children started coming frequently to Hunt's house in Mount Vernon. Then one day in 1957, after lunch with the two daughters from his first marriage, Hunt quietly disappeared to marry Ruth. He gave up gambling and started going to the Baptist Church. Becoming increasingly estranged from his children by his first wife, Hunt concentrated on work and writing, publishing ten books in as many years and writing continuous streams of letters to editors. He died, the symbol of both Texan ultra-conservatism and the Dallas super-rich, in 1974, aged 85.

The Second Generation

Hunt's immediate legacy was in the full soap-operatic tradition. Most of H.L.'s wealth had long since been salted away from the taxman in trusts and in Placid Oil and the Penrod Drilling Company, which was owned through trusts by the children of his first family. In his will, however, he left his house and stock in his personal oil company, Hunt Oil, to his widow Ruth with his son by her, Ray, as his executor. What he described as his Louisiana oil leases (his untied properties) he left in fourteen parts – six-fourteenths to go the six children of his first family, four-fourteenths to go to the children of his third family and the remaining four-fourteenths to go to the children of Mrs Lee. Each fourteenth was worth perhaps $100,000. The division, meant to be fair, had the effect of driving a final wedge between the various parts of the family. Nelson Bunker Hunt and Lyda's other children bitterly resented the apparent preference given to the third family in making Ray executor, despite the fact that they themselves already had most of the old man's wealth by then. Frania's family felt aggrieved because, although the children all had trusts, they had gained no share of the wealth distributed before H.L.'s death. In 1978 they brought suit in a trial that had all the family's dirty linen washed in public and was eventually settled out of court for $7·5 million.

Ray and his half-brothers never made it up. For weeks after the reading of the will, Ray telephoned his brothers who worked in the same building in Dallas as he did, the First National Building. His calls were never returned. Bunker and his brother Herbert were planning to form their own new company, the Hunt Energy Corporation, with some of the staff from Hunt Oil. Finally they received another of Ray's messages, marched straight up to see him and the two sides agreed to go their own ways.

Ray, called the 'Nice Hunt', by *Time* magazine, had so far concentrated most of his attention on real estate and art collecting. He now returned to the oil business and to taking up new leases. When the executives of an oil company came by the building to offer a share in their North Sea concession to Bunker Hunt, he was out so they did the deal with Ray instead. By the beginning of the eighties he had built up Hunt Oil to a company with holdings worth $1 billion and interests all over the United States, in oil, uranium and real estate.

Nelson Bunker and William Herbert meanwhile expanded even

more sensationally. By the time of their father's death they had already made Placid into a multi-billion-dollar company with a major oilfield in Libya and an important gas discovery in the North Sea. Their Libyan oilfield was nationalized in 1973 as a 'strong slap on the cool, arrogant face' of the United States for its backing of Israel. They also found themselves up in court on a far more serious charge of wiretapping some employees whom they suspected of theft. A criminal offence, wiretapping carried with it the very real threat of jail and it took a lot of political muscle and six years of legal pleading before the two brothers were finally found innocent in 1975 on the grounds that they were 'just plain folks' trying to catch the men who had been robbing their father, that they had no idea their actions were wrong and that their wealth should not be held against them.

The Scramble for Silver

Much less satisfactory was the outcome of the brothers' silver caper. From early on in the seventies Nelson Bunker Hunt seems to have become convinced that it was far wiser to invest in silver rather than in paper money which was rapidly depreciating. By 1974 he had already built up holdings of 55 million ounces of silver bought at less than $10 an ounce. He also started investing on the silver futures market. As the decade went on, his belief in seeking alternative forms of investment turned into an obsession, all the more so when heavy speculation in soyabeans brought him substantial profits. When even their enormous resources began to be strained by the scale of their borrowing, Nelson and Herbert sought Middle Eastern and other partners to buy with them, finally finding them in two Saudi sheikhs with connections, and possibly with additional financing, from the Saudi royal family. With renewed backing, the partnership, registered in Bermuda under the name International Metal Investment, purchased a further 90 million ounces, worth nearly $1 billion.

Unfortunately the old adage that what goes up must come down, applies to silver as much as oil. Worried that the Hunts were now in a position to squeeze the market, the commodity exchanges moved to introduce new rules early in 1980 limiting the amount any individual company could hold. This put the market into reverse as the traders sensed that the Hunts would have to start selling. At the same time the high price of silver had brought a stream of new supply as every householder searched his cupboards and attics for teapots and trinkets to sell. By March the price had fallen to nearly $20 an ounce and the more prices fell the more the Hunts had to put in to their accounts to show that they could meet any losses. On 25 March they were forced to admit that they couldn't manage it. Their financiers were stunned and when rumours of their difficulties leaked out, the price plunged all the faster collapsing from $15·80 to $10·80 an ounce in a single day.

As the Hunts had made many of their recent purchases at $35, their position was desperate. Although they still had huge stocks of actual silver, to try and sell it in a falling market could only mean still lower prices. Yet raising loans to finance their speculation was made difficult by the Federal Reserve's recent ruling that banks should not lend on commodity futures. The huge scale of the Hunts' investment and the consequent danger of a complete collapse not only of their companies but also of the financial institutions which had lent them the funds to buy futures and of the markets in which they had dealt, finally brought in the American Government in a highly contentious rescue package. A $1·1 billion loan was arranged by a consortium of banks. In a classic move the banks secured virtually all the Hunts' oil leases and production as well as their personal property, including that of Lamar who had been sucked into the operation only late on, and, of course, the silver. 'The terms of the loan were awful tough,' muttered Herbert.

The sight of the two Hunt brothers arraigned before a congressional committee to answer charges of trying to rig a market, just as the Seven Sisters had been seven years previously after the first oil shock, aroused deeply ambivalent feelings in the country at large. In one way, the junk-food-stuffed figure of Bunker Hunt, with his huge bulk and narrow eyes, alongside his greyer but well-groomed brother, Herbert, dodging questions and making sly comments to the camera, represented everything that was wrong with the riches of oil and the billionaires it had made. It was all too apposite that Texas wealth in this case should be allied with Saudi riches.

On the other hand there was something equally naive and 'American' about these two entrepreneurs from Texas, who'd gambled their all on one great 'scam' just like their father. 'Why would anyone want to sell silver to get dollars?' Nelson Bunker Hunt had exclaimed in sincere amazement as he was told of the famly silver being rushed to market at the height of the scare. 'I guess that they got tired of polishing it.'[5] Harry Hurt, no admirer of the family, none the less believes that the Hunts genuinely did not start buying silver to corner the market but because: '. . . they rather thought that paper money was no good; that it wasn't backed by real hard bullion any more and that this was the only way that they could stave off what they perceived to be at the time a coming apocalypse – the events of the Soviet invasion of Afghanistan, the hostage crisis in Iran and the general loss of US respect around the world during the Carter presidency.' The danger represented by the Hunts and their father, for Hurt, is precisely that they were 'hicks'. H. L. Hunt wasn't immoral but 'amoral, he played by his own rules not by everybody else's rules. In fact he thought that everybody else's rules simply didn't apply to him.'

The other side of the coin was the American folklore of the Texan independent and admiration for the man who would gamble everything on a well because he could sense there was oil there by the look of the land; the lone operator who could make his money because he was so much faster on his feet than the big, slow majors; the entrepreneur who could make his fortune because he had the courage to see things big and put his handshake on the deal.

Armand Hammer and Occidental

At one time during the first flush of their silver buying, when the Hunts also owned a sugar refinery, they put forward a deal under which they would sell large amounts of silver to the Philippines in exchange for raw sugar. The Philippines would in turn exchange the silver with the Saudis for oil and so the wheel would come full circle, the price of silver would rise, the Philippines would be able to sell their sugar and the Saudis their oil. It might have worked had the International Monetary Fund not intervened to say that they would not lend the Philippines money on the basis of silver.

It was precisely this sort of dealing energy, however, that made the fortunes of other oil billionaires like Paul Getty and Dr Armand Hammer. Hammer, who ironically made a $119-million profit out of the collapse of the silver price by selling short in preparation for the opening of a silver mine of his, came to oil both fairly late in his life and fairly late in the history of oil. He had arrived in Los Angeles in 1956 to settle down, already famous for a trading career begun by exchanging wheat for Russian furs and caviar under a deal struck with Lenin himself, when he was asked for some money to invest in a little oil company called Occidental Petroleum, whose shares had collapsed from $14 to 18 cents.

'They told me,' he recalls, 'that it was a good tax shelter to drill for oil, especially if you had a dry hole. You could write off your income tax. They offered me as many shares as I wanted but I looked up their balance sheet and said, "Your stock isn't worth 18 cents."

'"Well," they said, "we have two leases that we recently acquired and we'd like to drill them. We think there's oil in those leases and we'll take $50,000 to drill each well." So my wife and I loaned them the $50,000 and took a note in payment on the understanding that if we struck oil we could have a half-interest in those two leases and if we didn't we'd have a bad debt to write off. It was beginner's luck and both those wells came in.'

From that characteristic beginning, Hammer moved quickly into international oil, partly on the expectation of American imports, discovering oil in Libya in 1961 and coming into the North Sea in an equally independent way with Paul Getty, long a friend, and another friend, the Canadian Lord Thomson, then proprietor of *The Times* and the *Sunday Times*. Thomson was brought in because Hammer wanted a Scottish tinge to his application and Thomson owned Scottish Television and the *Scotsman* newspaper. Thomson agreed because he wanted big money and he trusted to Hammer's luck in drilling. He knew, he later recalled, that it must be a good deal when Getty took him to lunch, offered to buy his shares out and actually paid the bill. Hammer was lucky. Although a latecomer to the North Sea, he discovered the most profitable field of all in the Piper accumulation.

John Paul Getty

Getty, on the other hand, had started in oil as a child. His father, George Getty, an insurance lawyer, had gone to collect a bad debt in Oklahoma in the days of the oil boom and had bought a lease which made him a fortune. His son, John Paul, dropped out of college, spent his way round Euope on wine, women and song and returned to be offered a deal by his father to join in the business for a year and strike out on his own in search of oil leases. If he liked the business and made good, fine. If he didn't, the subject of his career would never be mentioned again. Young John Paul went to Tulsa and made his first million dealing in leases that others weren't interested in. After his father's death he expanded by buying companies such as Tidewater and Skelly Oil that were going cheap in the Depression. After the Second World War he looked, like others, to the Middle East, finally gaining the rights for the Saudi share of the Neutral Zone (so-called because a precise boundary could not be agreed) between Saudi Arabia and Kuwait by offering terms that gave the kingdom higher tax rates and more assurances of investment in schools, transport and hospitals than the majors, thus helping set off Saudi demands for more from the Amoco Consortium. Even so, it was nearly five years before he found oil, having drilled on through millions of dollars on the typical oilman's hunch that oil would be found. It was, in a 12 billion-barrel field.

Like H. L. Hunt, Getty was an odd mixture of womanizer, socializer and recluse. In 1957 it was his turn to be picked out as the richest man in the United States, this time by *Fortune* magazine. 'To my acute discomfort, the press had "discovered" me', he wrote in his autobiography, 'and I had become a curiosity, a sort of financial freak.'[6] As with Hunt, that curiosity applied as much to his habit of marriage – he had six wives in all – as to his business acumen. When he lay dying in 1976, cloistered in his Sutton Place Tudor mansion in the south of England where he had spent the last seventeen years of his life, fearful of the media attention in the United States and afraid of flying, the house was invaded by strident women seeking his final blessing and a place in his will. One even climbed through the window and was found rifling through his papers for signs of his bequest.

Again, like Hunt, Getty's will carefully divided his bequests between his various women on almost outrageously ungenerous terms. Like Hunt's will – dividing his shares in his oil company, worth around $2 billion, between his museum at Malibu (which also got his body for burial) and his grandchildren, to be held in trust for several decades – set off a torrent of law suits from various scions of his assorted families claiming ill treatment and sparking off quarrels between the family of far greater acrimony, greed and deviousness than any television soap opera.

The result, in Getty's case, was the establishment of his monument, the Getty Museum, the richest by far in the world, and the destruction of the one thing he had spent so much of his life in creating – his own oil company. The problem of his will was that, in leaving his majority shares in the company to both a trust for his grandchildren and his museum, he failed to regularize the relationship between these institutions and the management of the company itself. In the scrabbling for position and money that followed, the company and Gordon, his fourth son, fell out completely. Gordon, as the sole remaining trustee of the trust fund, fought to ally himself with the museum to get control over the company, while the company's management fought to ally themselves with other members of the family to reduce his control. The fight, which soon had the predators circling Getty Oil like sharks round a stricken swimmer, climaxed in a dramatic series of bids and counterbids for shares. Gordon Getty encouraged one bid, at $110 a share, from a Texan independent company, Pennzoil. The Getty management sought a 'white knight' to rescue the company and found it in Texaco, one of the Seven Sisters which John Paul Getty had spent most of his life fighting. Texaco offered $120 a share, later raised to $128 after the threat of legal action from members of the family, an offer, accepted by Gordon on the part of the trust and the museum.

Overnight the offer, valuing Getty Oil at $10 billion and thus making it the largest corporate bid in American history at the time in January 1984, doubled the value of the trust. It was a financial coup far greater in value than anything John Paul Getty had been able to achieve in his lifetime and was all the more satisfactory considering the oil price collapse that was to follow soon after. But it did what Getty himself could never have intended, killed off his company. In trying to avoid taxation and keep the wealth from being frittered away by his children, Getty achieved the opposite. 'Big Paul created this legacy,' sighed his daughter-in-law, Gail Getty, 'and what it's done, it's created unbelievable unhappiness. People have grown up who don't know how to cope with this, who they are, the name. If he did anything that's really wrong, that's it.'[7]

The legacy had an even more astonishing outcome a year later when Pennzoil took Texaco to court in 1985 to claim that the major had unfairly whisked its prize from beneath its nose after it had reached a gentleman's agreement with Gordon Getty. Texaco foolishly allowed the case to be heard in a Texan court, where the mere hint of a local independent being pushed aside by a major was enough to encourage a judgment of $11 billion against itself, enough, if not overturned in the appeal courts, to force the company into bankruptcy.

It was a sign of the shape of things that were coming to the American oil industry and the changing role of the independents. No longer drilling for oil, they are searching for shares. Not that dealing wasn't always as much a part of the game as wildcatting. In the myth it was the drilling that made the legend, the Dad Joiner character who was considered more real than the Hunts who bought him out. Even today there is a social difference in Dallas between 'old oil money', that is fortunes made before the East Texas oil strike in 1930, and those made subsequently. But the truth was always that Dad Joiner and his fellows spent more time in wheeling and dealing in leases than in working them. Getty, behind his daily barrage of telephone calls, got far more excited by conducting the financial deal than he did managing the business of drilling for, transporting, refining and selling oil. There is something in Dr Hammer's breathtaking schemes to establish fertilizer plants in the USSR, produce oil from shale in Colorado and develop the barter trade of China that thrills in the deal-making far more than in the projects' execution. Just as in the Depression of the 1930s, Hunt and Getty made some of their biggest coups buying shares and leases when the market was at rock bottom, so the fall in the price of oil in the mid-eighties has brought a new shake-up to the industry and a new kind of oil entrepreneur.

T. Boone Pickens

While Gordon Getty and his family were parlaying with Pennzoil and Texaco, another of the Seven Sisters, Gulf Oil, came under even more determined attack from a Texan independent, the colourfully named T. Boone Pickens. Pickens, the son of a Phillips Petroleum land agent, started off in the same company as a geologist. Now, in the mid-eighties he has established himself in the best Texan style as the populist defender of the small shareholder, the David of the industry bringing down the creaking Goliaths of the big oil companies, becoming one of the highest paid executives in the United States as president of Mesa Petroleum, in the process. In a series of daring and progressively ambitious attacks, Pickens launched bids for General American Oil, Cities Service Company, Superior Oil, Gulf and his old company, Phillips. He argued that the shares of each of the companies – all in the top 500 of *Fortune*'s biggest American companies' list – were grossly undervalued. In each case his bids pushed up the share prices of his prey, forcing them into desperate action either to buy him off or to push them into the arms of a more acceptable owner. So Cities Service went to Occidental, Gulf went to Socal, General American to Phillips. Superior Oil and Phillips both bought him off at the cost of incurring heavy debt to themselves. Pickens, while failing to come away with the prize of the company itself, has none the less come away with over $1 billion in dealing profit for himself and the stockholders of Mesa – over $500 million came from the Gulf deal alone. It is a record of which he is immensely proud and much of the traditional industry profoundly loathes him for it.

'On a short-term basis,' admits Ray Hunt, 'one cannot argue with the fact that he is making money for Mr Pickens and for those who stand side by side with him in his take-over activities. And there is some good in what he is doing in that he is causing the management of big companies to be very sensitive to how their stockholders are faring. However, that plus is far outweighed by the negatives. And the negatives are that the companies have to be able to engage in long-term planning. And if you are having to spend all your time and energy focusing on quarter-to-quarter earnings' results and almost week-to-week strategic legal tactics, that is terribly debilitating. It takes top management time away from where it ought to be spent and this is [on] long-term strategic planning.'

Nor is Pickens that popular among the other independents. 'I do not consider Boone Pickens to be part of the industry anymore,' says Michael T. Halbouty, the septuagenarian president of the Halbouty Oil & Gas Company in Houston. 'I consider him to be part of Wall Street.' 'He's got nothing to offer,' says Dr Hammer, who, like many other oil company heads in the United States, was forced to work out new defence tactics by Pickens' take-over ingenuity. The head of Cities Service, became so angry that he publicly announced that he was strapping on his six-gun and coming to get the man who had dared come after his company.

Pickens remains blithely unconcerned. Like old Hunt, he enjoys the notoriety, the pictures of him in the paper out hunting corporations as well as quail, and he relishes the role of the philosopher of the new small shareholder capitalism. 'Shareholders,' he declares, 'are the cornerstone of the free-enterprise system. If management ignores shareholders, the economic system that has allowed our nation to prosper will be jeopardized.'

The crux of his philosophy of oil is that the business has become what he calls 'self liquidating'. It is proving unable, at least in the United States, to replace the reserves it is using up. Not only are managements failing to find those reserves but, he argues, they have spent vast fortunes attempting to diversify out of the oil industry altogether into retailing, mining and other spheres, all at inflated prices.

'Major companies,' he says, 'are trapped. They have huge reserve bases to protect. But the opportunities to replace them through exploration are limited and expensive. The same is true in the North Sea and companies both in America and the UK which are not effectively replacing their reserves will either have to find ways to compensate shareholders or they will find themselves being taken over by stronger managements.' Parallel with this line of attack is his argument that the management themselves have a vested interest in going on spending and trying to keep their companies alive, even when their low share ratings suggest that they are grossly under-utilizing their assets. 'It's the empire syndrome,' he suggests. 'The managers look to their survival and the increasing size of the company. Most of them don't own many shares themselves. They're not interested in the stockholder.'

The key to Pickens' success has been the simple observation that with falling oil prices and rising costs in finding new oil, it is 'cheaper to buy oil on Wall Street than go out and look for it yourself'. The means of his success has been his ability to raise money on the basis of the assets he is attempting to purchase. It costs him very little to launch the bids. He simply gets the support of a group of banks, or issues bonds, on the basis that the loan won't actually be drawn down unless the bid is successful. The object is to make money. The rationalization behind it is that it gives the shareholder a square deal. 'I was once warned by a friend,' he says, 'not to be a crusader. "They don't make money," he told me. "Well I don't want to be no crusader then," I retorted. "I want to make money."'

'When I make a bid,' he comments, 'I don't mind which way it goes, whether I take over the company or get bought off. The important thing is that the shareholder gets a better deal. I made nearly $600 million for Mesa's shareholders in the first four deals. But I also made a total profit of $8,768 billion for the shareholders of Cities Service General American and Gulf. And that's money in people's pockets to spend. And, I might add, it's $3 billion in federal taxes.'

The blatant air of profiteering through share raids, 'greenmail' as the process of bidding for companies with borrowed money in order to be bought off is called, is proving profoundly offensive to many as companies have been forced to seek more and more expensive and complicated ways to defend themselves. But it is the suggestion that the domestic American oil industry is an industry on the way out that has really angered so many of his compatriots in Texas.

'Boone Pickens has a negative philosophy,' complains Michael T. Halbouty. 'He doesn't believe in the industry. He calls it a sunset industry. He's even made statements that have been published that finding new reserves is almost impossible. That's his philosophy. Mine is of optimism. I believe there is a very high potential for finding new reserves in this country. No country, no company can grow without more exploration. If you stop completely what's going to happen to these companies when they deplete their reserves? Then they won't have any money to do any more exploration. He is so totally wrong. He claims that he is a geologist but for him to be a geologist and state that there is no more potential in this country, then he has lost his geological attributions somewhere down the line.'

The 'greenmail' aspect is bringing its own reaction, just as the Hunts found in their silver-buying spree. In the spring of 1985 Pickens suffered his first real reverse when his new prey, Union Oil of California, took to the courts and won its case, forcing a drop in share price that will give Pickens a substantial loss. At the same time the Federal Reserve and the Securities and Exchange Commission started to plan new rules to limit 'leveraged' buy-outs, bids based on loans raised against the break-up value of the company being taken over. None the less, Pickens' fundamental points are posing more awkward questions. There are many who sympathize with his view that oil company managements have ignored the shareholders in their own pursuit of empire, that the industry's record of diversification – Exxon with electronics, Shell with nuclear power, Mobil with retailing and Sohio with copper – has been abysmal and that, in an age of oil surplus, the industry will have to be restructured from top to bottom, the big even more than the small.

'The long-term argument,' he declares, 'is meaningless when companies are facing continuous decline in their domestic reserve base. The short term is the only thing to consider. There are other companies than Gulf in similar circumstances where their assets are worth more than their share price. As long as that exists, I believe we'll see more merger and acquisition activity. All they are waiting for is a catalyst.'

Pickens clearly sees himself as that catalyst, the smart Texan in a city suit ready to beat the big boys at their own game. But to suggest in the sunrise belt, the land that was made by oil, that oil is a sunset industry? That takes gall.

Above: An 1895 portrait by Eastman Johnson of John D. Rockefeller, the giant of the early years of oil, whose Standard Oil Trust became the greatest corporation in the United States, a symbol of industrial power and competitive ruthlessness.

Overleaf: A cartoon depicting the San Remo carve-up of the Middle East between Britain and France in 1920. The United States forces her way into the discussion, demanding an 'open door' policy for her companies also.

TURKEY
SYRIA
JORDAN
IRAQ
SAUDI ARABIA
TREATY
CONDOMINIUM

AGREEMENT
DECLARATION
PATENT

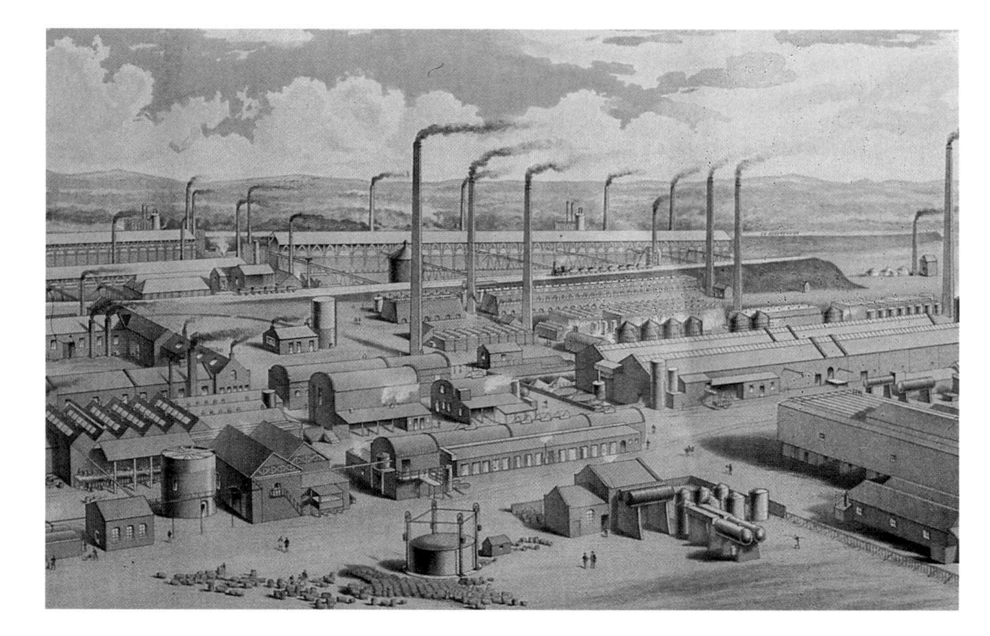

Above: A painting of 1900 depicting James Young's Paraffin Light and Mineral Oil Company in Glasgow. Incorporated in 1866, it was the world's first oil refinery. Young patented the process for converting oil from shale into lighting fuel in 1851.

Below: The earliest drilling methods employed by oil explorers. 'Colonel' Drake was the first to use the drill, hammering it into the earth at 'Drake's folly', in 1859. Later developments introduced the rotary drill, which screwed into the earth.

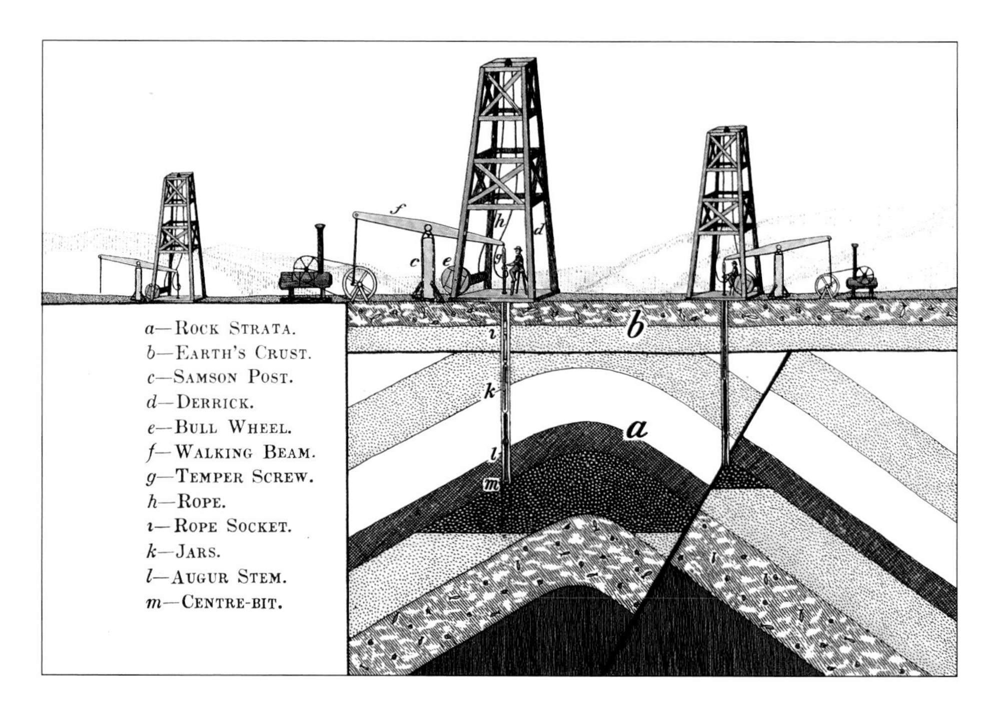

SHELL
R THE UTMOST
HORSE POWER

Happy motoring. The motor car revolutionized the oil industry. *Above:* A Shell poster of the 1920s (*left*) and (*right*) a topical cover from *L'Illustration* in 1923 of a traffic jam after a car rally. *Below:* A garage in 1926. At first petrol was sold in cans as was paraffin. Garages and petrol pumps only became widespread during the 1920s.

A semi-submersible drilling rig working in the North Sea. The rigs, supported by huge pontoons under the waterline and anchored to the seabed when drilling, enabled the industry to drill in up to 304 metres (1000 feet) of water and surmount waves of 30.4 metres (100 feet) and more.

Above: Working on the drilling platform. Offshore workers, or 'roughnecks', add a 9.1-metre (30-foot) section to the drill pipe which can bore 4 miles into the earth's crust.

Below: Much offshore maintenance work is done by divers. Here, a section of underwater pipe is photographed with specialized equipment designed to show up any cracks in the welds.

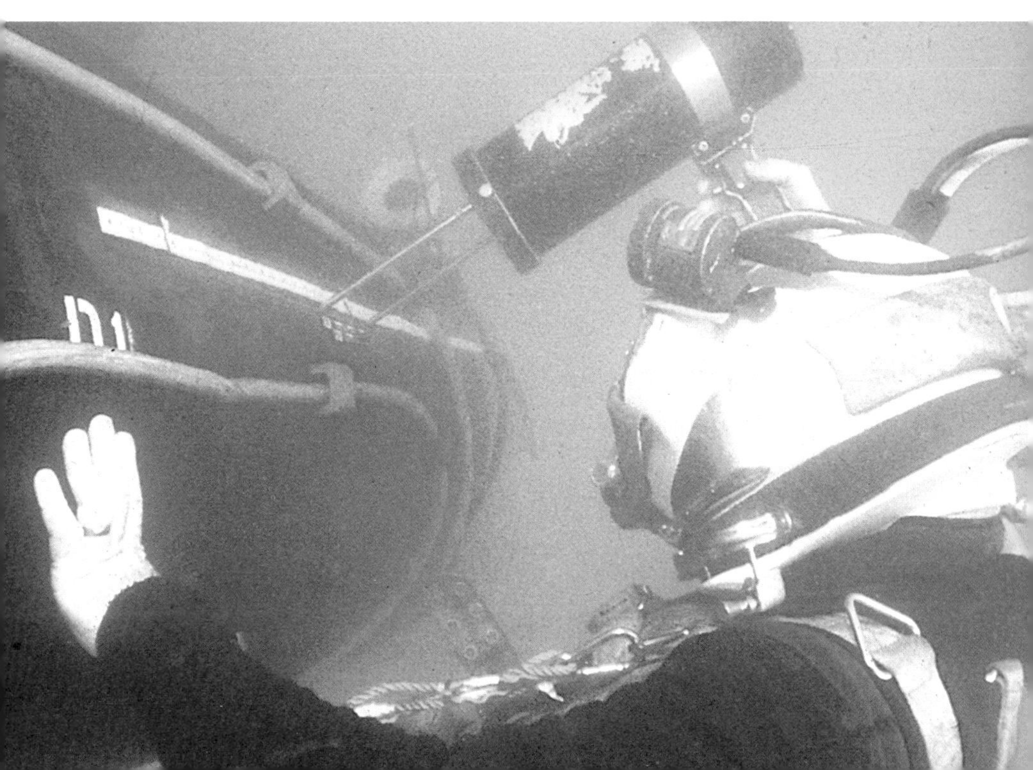

Alternative energy. *Above:* Solar power panels in Saudi Arabia, an infinitely renewable energy source and, even for the Saudis, the fuel of the future for the developing world. *Below:* The 'charge face' in the reactor hall of a British nuclear power station, the heart of the plant where the reactor heats water under a controlled process of fission.

6

'The Devil Gave Us Oil'

As the price of oil soared in the 1970s, all eyes turned to the Arabs and their new wealth. As the price started to fall in the eighties, the concentration switched to the oil debtors and to Mexico in particular. In August 1982, the then Finance Minister of the government in Mexico, Jesus 'Chucho' Silva-Herzog, flew to New York to meet the top representatives of his country's major creditors. Mexico, he declared, was technically bankrupt. Its revenues were exhausted. Not only was the country unable to repay any of its huge foreign debts, then totalling some $65 billion, but it would need as much as $6 billion more even to meet its interest payments. The country would have to renegotiate, or. . . . He left the second part of the sentence hanging, but his audience knew what he meant. If Mexico reneged, so too would every other Latin American borrower and every other oil-exporting country such as Nigeria which had stacked up immense foreign borrowing during the years of oil plenty. The 'Debt Crisis' and the massed effort of banks, governments and international agencies to sort it out, had been set in motion by its most spectacular victim.

Mexico's position was particularly tragic. For, from the start of oil development there in the opening years of the twentieth century, the country had always been deeply ambivalent about whether the costs of developing an indigenous oil industry would outweigh the benefits. 'At the beginning of the century,' says Herzog, with an air of slight fatalism, 'there was a Mexican poet who wrote that, "the fields of corn were given to Mexico by God and wells of oil by the Devil". We are still fighting,' he adds, 'to see if the poet was right or if the poet was wrong.'

Unlike Arabia and the Middle East, where oil had been the first industry to bring development to the desert emptiness, when oil came to Mexico the country was already deeply scarred by internal conflict between landowner and peasant and by resentment at foreign exploitation of its mines and infant industry. Mexico has always seemed unable to break loose either from its own turbulent history or its position beneath the might of the 'Yanqui'. Far from easing the tensions, oil simply exacerbated them.

Edward L. Doheny

Even at the outset, the first discoveries of oil in Mexico were the object of an intense struggle for influence between Britain and the United States. From the United States came Edward L. Doheny. One of those near-legendary figures of the American southwest, Doheny had been a prospector for gold and silver, who had served for a time as a chief vigilante in a New Mexican town, and, after a run of bad luck, had fetched up as a barrel-cart driver for an oil company in California. Seizing the opportunity, like so many of his generation, he moved with a small bundle of savings into the oil-lease business. In good prospecting style he concentrated his attention on what could be seen on the surface, in this case tar pits and heavy oil seepages. Instead of drilling, he started digging. He dug out enough to make a little money with which to buy more leases and so he spread his net around California to wherever he saw signs of the sticky black liquid. For most of the early explorers, the goal was clearly light oil which could be easily refined into kerosene. Doheny, following the example of Lyman Stewart, founder of Union Oil of California in Los Angeles, sold his heavier product to pave streets and to burn under boilers and railway engines.

It was railway engines that brought him to the attention of A. A. Robinson, one of the directors of the Santa Fé Railroad. Moving to Mexico to take charge of the Mexican Central Railroad, Robinson invited Doheny down in 1900 to find fuel for the engines. Doheny and his partner, Charlie Canfield, came in a private carriage and set out to explore the swamplands of Tampico in southern Mexico. After months of searching they finally came across a clearing in which they found a hill springing oil from its sides. As Doheny wrote later:

> The sight caused us to forget all about the dreaded climate, its hot, humid atmosphere, its apparently incessant rains, those jungle pests the *garrapatas* [sheep ticks], the dense forest jungle, which seems to grow up as fast as cut down, its great distance from any center that we could call civilisation and still greater distance from a source of supplies of oil-well materials – all were forgotten in the joy of discovery with which we contemplated this little hill from whose base flowed oil in various directions. We felt that we knew, and we did know, that we were in an oil region which would produce in unlimited quantities that for which the world had the greatest need – oil fuel.[1]

Weetman Pearson

At almost the same time Weetman Pearson appeared on the Mexican oil scene from Britain. Head of the family contracting firm of S. Pearson and Sons, one of the world's leading tunnel- and dock-construction companies, Pearson had come to Mexico in 1889. Over the following years his firm drained the Mexico City valley, built a new harbour at Vera Cruz and constructed a 200-mile railway across the Tehuantepec Isthmus between the Pacific and the Gulf of Mexico.

It was while looking for quarry stone for the port being built at the

Gulf end of this line that one of Pearson's partners, J. B. Brody, discovered oil seepages, just as Doheny had. At first Pearson took little notice of this until he found himself spending the night in Laredo, Texas in April 1901 after missing a train connection. It was at the time of the great Spindletop discovery and the whole of Texas was abuzz with the news. Pearson made further inquiries and then, gambling on the assumption that if there was oil on the north side of the Gulf, there might be on the south too, he instructed Brody to buy up all the land he could around his discovery.

Neither he nor Doheny were immediately successful. Doheny, who sank his first well in 1901, spent over $3 million and nearly ten years of effort in producing just small quantities of heavy oil, most of it only good for street paving, before hitting the big time. Pearson, who had bought his first concession in 1902, spent £5 million and pledged most of his fortune in an ambitious strategy to build an entire industry of pipelines, refinery facilities and tankers before he found the oil to sustain his enterprise and meet his sales commitments. He was not known for the 'Pearson luck' for nothing. In 1906 his friendship with the 76-year-old dictator, Porfirio Diaz, helped him gain a huge new concession to the north of his original lands. It was on this that, in 1908, he brought in his first gusher at Dos Bocas.

The well blew with such force that it took all the wellhead equipment with it, caught fire and burned for eight weeks, destroying a million tonnes of oil and vast areas of grazing acreage. Two years later, with the help of a team of young geologists recruited from the United States, who included Everett deGolyer, Pearson's company brought in the Petrero del Llano No 4, the largest gusher of its time. It flowed oil at a rate of over 100,000 barrels per day. It also unfortunately flowed 'wild', setting loose a tidal wave of black oil that engulfed the farmland between the site and the sea and then polluted several hundred miles of coastline before being capped. In the same year Doheny brought in his first big gusher, the Cerro Azul, in an adjoining area to his first find. At 70,000 barrels per day it also came in 'wild' with devastating results on the surrounding countryside.

This succession of discoveries made Mexico the second largest oil producer in the world, with an output of 4 million tonnes by 1914, and it enabled Pearson, who controlled some 60 per cent of it, to build up a worldwide network of sales as far as Russia. It also gave rise to an oil boom of unparalleled violence with a ruthless struggle for market dominance between Pearson and Standard Oil, leading to the eventual downfall of Diaz.

The Struggle for Dominance

Mexico was not a land of half measures and it gave scant consideration to the local Indian *peon* or peasant, who was generally regarded as something less than quite human both by the foreigners coming in to make money and the resident landowners tracing their heritage back to pure Spanish blood. As the first indications of oil attracted the attention of oil companies and prospectors, land was snapped up with merciless energy and avarice. The bribing of judges and local officials was common. When bribery did not serve, even more extreme tactics were employed and it was not unknown for peasants to be shot and their thumbprints attached to a bill of sale. In a land too ignorant to understand what was going on, Doheny was able to buy up some 450,000 acres within a few years, while Pearson eventually owned 600,000 acres and oil-prospecting leases covering an area of 1 million acres.

'It's got the Klondike faded,' an old prospector told the writer Jack London. 'We're taking $90,000 a day in gold from an eight-inch hole in the ground.' London, who had been sent down to cover the new Klondike for an American magazine, helped create the image of violent exploitation which profoundly affected liberal American public opinion as well as Mexican folklore for years to come. 'The oil has spoiled the land . . .' wrote the American, Frank Hanighen, in the 1930s, 'but the foreigners have no time to contemplate the misery they have wrought. They are absorbed with the struggle of their drills against the earth, and with the struggle among themselves.'[2]

The major 'struggle between themselves' concerned Pearson, who had reorganized his holdings into the Mexican Eagle Petroleum Company with Diaz's son as a director, and Waters-Pierce, an American company, two-thirds owned by Standard Oil, to whom Doheny now sold all his oil. When Pearson decided to sell his refined products into the domestic market in 1908, Waters-Pierce reacted furiously. A prolonged oil war broke out between the two, involving sabotage, banditry and kidnapping as well as price cutting until a truce was made with Standard Oil in 1911.

By this time the thirty-year rule of Pearson's main ally in Mexico, President Diaz, was coming to an end. In the country the authority of his army was being progressively challenged by a number of local peasant armies led by guerilla leaders like Emiliano Zapata, Orozco and Francisco 'Pancho' Villa. In Mexico City itself the middle classes and the intellectuals were finding a new voice in Francisco I. Madero, who was widely admired in the United States and partly funded by Standard Oil. Diaz fell in 1911, slipping off to a comfortable retirement in Paris, while Madero was installed as his successor amid great rejoicing. Idealistic, intellectual and the son of one of the richest families in Mexico, Madero was unable to control the rebel armies, who demanded land reform and refused to disband. Nor did he manage to play off successfully the competing aims of Britain, the United States and the oil giants that soon engulfed him. With the discreet backing of a number of commercial interests, General Victoriano Huerta intervened as the strong man. Declaring that he was there to bring order to Mexico in the face of chaos, he arrested Madero 'for his own protection' and had him assassinated while being moved to prison.

Order was not what Huerta brought. The rebel armies continued to swarm over the countryside and over the oilfields, where mercenaries had to be employed to protect them. Britain recognized the new regime but the United States, under a new and liberal President, Woodrow Wilson, did not. American public opinion was outraged by Madero's murder and the newspapers were soon in full cry against the British and Pearson (now Lord Cowdray). Doheny himself appeared before a congressional hearing to plead the case that 'foreign' oil interests had to be stopped if they were not to monopolize non-American oil supplies – a charge given added force when Cowdray succeeded in signing a substantial fuel-oil deal for the British Navy in 1913.

As so often in Mexican history, the struggle for power within the country became the focus of competing outside interests. The British and Cowdray continued to support Huerta. Standard Oil funded the opposition of General Venustiano Carranza to the tune of $\$\frac{3}{4}$ million in cash and credits. In March 1914 President Wilson used the arrest of an American naval officer in Vera Cruz as a reason for sending in American troops. Urged on by the Hearst chain of newspapers and a wave of jingoism, the American intervention achieved its objective. Huerta was sent into exile in Spain and his place was taken by Carranza.

The new President, however, proved to be less amenable than the Americans had at first thought. Having been brought to power by the 'Yanquis', he was doubly determined to prove his independence of them and their commercial interests. As part of the new Mexican constitution, Article 27 was inserted, stating that 'in the Nation is vested the legal ownership of all minerals, petroleum, and all hydrocarbons – solid, liquid or gaseous'. Only Mexican individuals and companies would be allowed concessions or ownership of such resources. If foreigners were to gain rights, they would have to establish locally controlled subsidiaries and 'accordingly not invoke the protection of their governments'.

The first reaction of the oil companies was to drill for all they were worth and produce every possible drop of oil before the constitution took effect. The next was a breach in relations with the United States, caused by a number of incidents leading up to a raid by Pancho Villa across the border into New Mexico which brought an ill-fated and largely ineffective counter-invasion by General John Pershing. Finally, after a prolonged period of upheaval within the country, Carranza himself became the victim of a coup and was shot dead by one of his own officers.

The Pioneers Exit

By this time both Cowdray and Doheny had sold out. Pearson, increasingly disturbed by events within Mexico, had been having discussions as early as 1911 with Texaco and Gulf but the talks had fallen through for anti-trust reasons. In 1912 Calouste Gulbenkian had tried to get him together with Shell and in 1913 he had met with a director of Standard Oil (New Jersey). None of these talks came to anything and during the First World War Lord Cowdray had thrown himself into much more ambitious plans to form a new imperial consortium around Anglo-Persian and Burmah Oil. When this also failed, he was approached immediately at the war's end in October 1918 by Gulbenkian once again, acting for Henri Deterding. Shell bought into Mexican Eagle early in 1919. Cowdray had been lucky. Within three years the main producing fields had started to bring up salt water instead of oil.[3]

Doheny sold out his company, the Pan-American Petroleum and Transport Company (which was part of the Turkish Petroleum Company at the time), to Standard Oil of Indiana in the twenties following the Teapot Dome scandal. The scandal, the worst to darken American history since the Administration of General Grant fifty years before, had involved bribing the American Government to release land held by the US Navy in case of emergency for the exclusive exploration and development of a few select companies. The man bribed was the Secretary of the Interior, Albert Fall. The oilmen after the concessions were Harry Sinclair and Edward Doheny.

Sinclair was a larger-than-life Oklahoma independent with larger-than-life ambitions. (He had tried to negotiate with the Russians for concessions in northern Persia and Russia during the early years of the Revolution and the Sinclair Oil Company had originally also been part of the Turkish Petroleum consortium.) When Woodrow Wilson was replaced by President Harding in 1918, Sinclair and Doheny moved smartly to press for the opening up of oil leases held in reserve for the US Navy. With the help of Theodore Roosevelt, Assistant Secretary of the Navy and a former director of Sinclair Oil, and Interior Secretary Fall, a former senator of New Mexico and a friend of Doheny, the lands of Elk Hills in California and the Teapot Dome in Wyoming were transferred from the control of the Navy Department to the Interior Department. They were then leased, in 1922, to Doheny, who got Elk Hills, and Sinclair, who got Teapot Dome, on an exclusive basis.

The two men declared that they hoped to make a $100 million out of their concessions. Secretary Fall appears to have benefited to the tune of $400,000. All seemed to be going well – Albert Fall had resigned to comfortable retirement in New Mexico and Sinclair was busy pursuing his Russian ambitions – when President Harding died of food poisoning. The lid came off the oil can – with a little bit of prodding, it was said, by Standard Oil – under Senate investigation. After five years of court hearings, Fall, Doheny and Sinclair were convicted of bribery and sentenced to prison in addition to being fined. Doheny and Sinclair were both released on technicalities. Fall's sentence was upheld, leaving the ironic implication that it was all right to offer bribes but not to receive them.

Doheny went on to die a rich man. Like Lord Cowdray, he was fortunate in getting out of Mexico when he did since oil became an increasingly contentious part of Mexican politics during the inter-war years. Carranza was replaced by a triumvirate of three: the intellectual Adolfo de la Huerta, who wanted to use the British to counterbalance the Americans; the tough-minded General Alvaro Obregon and the labour leader, Plutarco Elias Calles.

Article 27

It was not to last. Although the British and Americans and the various commercial interests continued to try and back the winning side in the struggles between Huerta and his colleagues which followed, it was soon apparent that Mexican political feeling was becoming increasingly anti-American and anti-oil. Almost as soon as General Obregon became President in 1920 he raised the taxes on oil and announced his intention of implementing Article 27 of the constitution. Strong objection from the Washington Government ensured an uneasy alleviation of the commercial tensions when Plutarco Calles succeeded as President. Calles, however, owed much of his support to the labour organization and to the oil workers. The oilfields were hit by a series of strikes, lockouts and sabotage and the cry of 'twenty-seven' was taken up as the slogan of the labour unions. Calles, wanting to strengthen his position at home, decided to impose Article 27 on the companies to accusations of 'expropriation' and strong protests about proper compensation and legal rights of contract from the pro-oil American government. Article 27 aroused much the same high public indignation in the United States as the nationalization of Anglo-Iranian did in Britain thirty years later. Mexico, however, was on the United States' doorstep and public opinion was further inflamed by Calles' crackdown on the Catholic Church.

Mexico's oil revenues, which funded a third of its national budget, were cut by a third and exports slashed by even more. The American newspapers were demanding action when the Washington Government tried one more effort at conciliation through the appointment of Dwight Morrow, a partner in the New York investment house of Morgans as Ambassador. One of the most skilled diplomats of his time, Morrow felt that Calles, desperate for foreign capital and an easing of his oil problems, could be treated with. Relations between the two countries eased. A Supreme Court ruling stopped the Mexican Government from refusing regular working permits to companies resisting the application of Article 27. The shape of political power was changing in Mexico, however, not least because of the influence of organized labour and its influence in the ruling party. The truce between the Government and the oil companies lasted only until 1934, when the Depression saw the election of the incorruptible socialist Lazaro Cárdenas, as President.

At first Cárdenas shied away from the oil sector as he started to nationalize the mines and factories. But in 1937 the Oil Workers' Syndicate went on strike, demanding worker participation in management, shorter hours, the extension of union benefits to families and a substantial pay increase. The oil companies refused. The Mexican Supreme Court upheld an arbitration decision in favour of the workers and, in March of the following year, the Government declared the expropriation of the industry and then proceeded to send in the Army to take over all oil facilities. Mexico had moved where even the more radical countries such as Venezuela feared to tread for at least another generation.

The immediate result, as it was later in Iran and Libya; was the effective isolation of the country as the oil companies went to the international courts, backed by their respective governments. Mexico found itself able to sell only to the Fascist powers of Germany, Italy and Japan and then only on a barter basis. During the war its exports virtually ceased and it was not until the end of the war, with a new President, Manuel Avila Comacho, that compensation terms were finally reached with the oil companies amounting to a total of some $50 million.

Over the ensuing decades Mexico took cautious steps towards re-entering the international energy market. It held tightly to its belief in national control and the dominance of all oil affairs by its state oil company, Pemex, founded in 1938. Within Pemex and in the broader political field, power came to rest increasingly with the industry's unions. Despite the reintroduction of the international oil companies as explorers under service agreements with Pemex during the 1950s and 1960s, oil development slowed. Many of the older fields were exhausted, in some cases because of earlier over-exploitation. Pemex was anxious to move at a pace with which its own technology could cope. Mexican production stabilized at around 20–25 million tonnes a year (500,000 barrels per day), leaving Venezuela to feed the United States' growing import requirements. The legacy of fifty years of foreign commercial domination had given oil a bitter image within Mexico and the country preferred to concentrate on gas, discovered in large quantities on the United States' border just after the war. Mexico, once the second largest exporter in the world, was not to return to the market until the mid-seventies.

Pemex and Going for Growth

The change came with the first oil shock of 1973–4. By then, despite the efforts of Pemex, demand within Mexico was fast outstripping its oil-production capacity and the country seemed destined to become a major oil importer when rumours started to spread in the American contracting industry that the state oil company had made a major new discovery. At first these were dismissed as wishful thinking. It had been so long since the majors had been involved in Mexico that knowledge of the country's geological potential was surprisingly limited. A series of discoveries in the southeast during the following years, however, rapidly confirmed that Mexico had indeed established a new oil province that was of worldwide significance, just as the world was desperately seeking new sources of supply outside the Middle East.

With the election of Lopez Portillo as President in 1976, and encouraged by an increasingly self-confident and ambitious Pemex, Mexico embarked on a dramatic go-for-growth policy based on oil. Funded by outside capital, eagerly lent by the commercial banks of the United States, Pemex threw itself into a major programme of expansion. Between 1976 and 1982, the six years of Portillo's rule, production increased fourfold from 800,000 to 3 million barrels per day, while exports were pushed up from virtually nothing to 1·5 million barrels per day in 1982, making Mexico one of the top five oil exporters in the world.

Each year Pemex announced substantial increases in its proven reserves of oil. When Portillo came to power, recoverable reserves of oil and gas in Mexico were calculated at just over 10 billion barrels of oil equivalent. By the end of his term they were officially declared at 72 billion. As the exports rose, Mexico increased its borrowing from the international banks, which were all too eager to get a share of the action in a country which seemed destined to have one of the fastest growing economies in the world in the post oil-crisis era. And each year, it was true, Mexico recorded growth figures double that of the world average. Mexicans, declared Portillo proudly, would now have to live with the challenge of 'administering abundance'.

Those problems proved far greater than Mexico and its bankers ever suspected. The oil sector fulfilled its targets but the programme of 'grow now, pay later' imposed intolerable strains on the nation's economy. Just as with the Shah of Iran's similarly grandiose plans for industrial expansion, the speed of development was more than the infrastructure could cope with. Wages of the 150,000 oil workers rose out of all proportion with those of other employees in industry and agriculture. Inflation soared into double figures and then nearly into triple figures as it touched 90 per cent in 1981–2. Corruption was rife. Over-hasty industrialization brought with it all the scars and tragedies of pollution and waste. Foreign debt more than doubled from $23 billion at the end of 1977 to over $60 billion five years later.

The 'Debt Bomb'

It couldn't last and it didn't. Interest rates suddenly spurted in the aftermath of the second oil shock in 1979–80. Over half the country's foreign revenues were now needed just to cover the interest on its foreign debt. Yet the main source of those earnings, oil, was beginning to fall in price. In 1981, Pemex, which had been charging a premium for its oil, was forced to reduce prices to keep to its target of selling 1·5 million barrels per day. Political reaction to the move from the unions and party members was such that Pemex's chief, Jorge Diaz Serrano, was hurriedly transferred to a less exposed position as Ambassador to the USSR. Portillo borrowed nearly $10 billion more in 1981 just to cover his interest payments. Yet, as much out of pride as of reason, he refused to devalue the Mexican currency. Private money responded to the obvious signals that the Government was ignoring. Within a period of just eighteen months between 1981 and mid-1982 some $40 billion of Mexican capital flowed out of the country to safe havens abroad, much of it transferred by government officials and ministers.

There began what the then Finance Minister, Jesus Silva-Herzog refers to as the 'days of madness'. While Herzog and his officials flew to New York to plead with the international bankers for a renegotiation of Mexico's debt and the country awaited the new presidency of Miguel de la Madrid Hurtado, picked as Portillo's successor but yet to be formally installed, Portillo called a crisis Cabinet meeting and, as a final act, announced the nationalization of the banks, the imposition of exchange controls and a call to arms with his fellow Latin Americans to renege jointly on their debts.

It wasn't a course which Mexico could follow, at least not in the middle of a change in government. De la Madrid Hurtado called a halt as soon as he was in office in December 1982. Herzog and his officials were despatched once again to New York, as well as to London and Tokyo to seek a rescheduling of the debt. They succeeded but only with the help of the central bankers of Britain and the United States, both of whom were terrified at the prospect of a general strike by the major debtors that would bring down most of the leading commercial banks of the United States and Europe with it. It was also achieved at the cost of some humiliating pleading. 'I will always remember the president of one big US bank,' recalls Enrique Castro, a Mexican official who took part in the negotiations, 'which only a year before had been pushing us to borrow money, sitting on the other side of the table puffing on a cigar.

' "Mr Castro," he said, "one thing that you will learn is that life is hard when you want to reschedule your debts."

' "But, sir," I exclaimed, "you cannot impose these conditions. No other country in the world would be asked such stiff terms."

' "That's my point, Mr Castro," he replied, taking another puff of his cigar, "rescheduling comes expensive."

'They charged us,' Castro adds, 'for everything – for the baby-sitters that they said they had to employ while we negotiated late in the night, the theatre tickets for the shows they had to miss, for the taxis they kept running outside and forgot to let go.'[4]

When the terms were finally settled, Mexico got its money. A total of $20 billion of its debts was rescheduled over five years and $5 billion in new loans was promised to see it through the following year. But the conditions, agreed with the International Monetary Fund (IMF), were indeed harsh: the devaluation of its currency, a reduction in government expenditure and a lessening in government subsidies on food and clothing, as well as a tight control of wages. As a Mexican banker has commented: 'Our recent history is one of perhaps the biggest failure of any country in the contemporary world.'

That 'biggest failure', or 'debt bomb', as it was quickly termed in banking circles, raised two fundamental questions both for Mexico and for the world of oil as a whole. One was whether Mexico had been right at all to restart exports in 1976 after a gap of nearly forty years. In the eyes of critics the decision to go for petroleum growth proved little short of an unmitigated disaster. 'Petroleum,' argues Alberto Castillo, Professor of Mathematics, head of the Mexican Workers' Party and one of the regime's most persistent opponents, 'for a poor under-developed country like Mexico is a blessing. But only if this petrol is kept within its national confines and is not sold abroad. Petroleum produces riches where it is consumed and not where it is produced. The money that reaches Mexico through the export of petroleum goes out again to buy manufactured goods. If one follows the trail of a barrel of oil, it leaves the poor country and it reaches the industrialized country to produce the goods which later on return to us at an exorbitant price. Always in the sale of crude petroleum, you lose something.

'Every day Mexico is paying back interest equivalent to 1·25 million barrels of oil. Such is our luck that if we continue to sell, and we continue with our loan, we're never going to get out of the problem. Then, in order to resolve the situation, we have to go back to 1976: Mexico didn't sell petrol abroad, and Mexico was a nation that was growing at the rate of 6 per cent annually.'

That is not a view that the Mexican Government accepts. Mario Ramon Beteta, the financier who was put in to head the reorganization of Pemex in 1981, argues that: . . . 'the type of investment that was carried out by Petroleos Mexicanos [Pemex] in huge amounts enabled this institution to increase its production from 800,000 barrels per day to almost 3 million barrels per day in a very short period of not more than seven years. This is quite an achievement by any standards in the world.' But Beteta, like his colleagues in de la Madrid Hurtado's governments, would accept that Mexico, like other countries, believed too easily that '... demand would go on increasing for ever and so, therefore, would prices. Now that prices are going down and demand is shrinking, it is necessary to manage oil activities more carefully, in a more austere way.'

For Pemex this meant not only a slashing of expenditure and a sharp reduction in the foreign loans of nearly $20 billion that had been incurred during the heavy days of expansion, but also some pruning of the extensive corruption, of both the management and the oil unions, for which it had become notorious in the seventies. One of de la Madrid Hurtado's election slogans had been the need for 'moral renovation'. Within a few months both the Chief of Police, who had built several palaces and amassed a considerable art collection in the previous decade, and the former head of Pemex, Jorge Diaz Serrano, who was accused of embezzling $34 million, were arrested together with a number of union leaders. Critics, such as Castillo, however, maintain that they are just the tip of the iceberg. 'The Petroleum Union,' he comments, 'is the most corrupt thing in this country. I'd say that it was the nest of corruption. And its leaders are a bunch of real gangsters. They are guilty of offences that the voice of the people has cried out against, for robberies, kidnapping and assassinations, and they have been sanctioned by the President of the Republic and praised on television.'

The problems lie partly in the old argument of idealism versus pragmatism, given particular force in Mexico by the degree to which the unions have been brought into the political system over the past fifty years. It also, however, gives rise to a newer and persistent questioning of the role of oil in the industrial development within the Third World. President Portillo, much influenced by British Keynesian economists, saw oil as the key with which the State could drive Mexico into industrial take-off, directed from the centre. Critics, from both the right and left, see it as intensifying the imbalances between rich and poor, between agriculture and industry, the land and the city.

Mexico City, which, in the post-war years, became the focus of industrial investment and federal spending, saw its population increase tenfold from 1·8 million in 1940 to over 18 million in 1982, making it the most populous city in the world, with more inhabitants than even Tokyo and Yokohama. Even in the post 1983 recession, its numbers are swelled by the arrival of around 1000 peasants from the country each day, scratching a living in the shanty towns and slums that have sprung up all around the city on every available piece of spare land. By the end of the seventies Mexico City had the unenviable reputation of being not only the most populous but also the most polluted city in the world. The San Juan Ixhuatepec gas explosion disaster in November 1984, when a leaking pipe caused a series of Pemex gas storage tanks to explode killing nearly 600 and seriously injuring 2000 people in the vicinity, appeared as a dreadful example of what uncontrolled industrial development within a city could do. The earthquake of September 1985, when some 700 buildings were wrecked, over 5000 were killed and 50,000 left without work, proved a still more terrible blow.

It also raised the second major question that overhangs Mexico still – will the medicine that it is taking to overcome its problems actually work? The pill has been bitter, as Jesus Silva-Herzog, who was dropped for his alleged subservience to foreign bankers' demands in 1986, admitted: lower expenditure, limited subsidies on food and clothing for people already living below the bread line, reduced wages and foreign exchange restrictions. 'We are fighting an economic crisis that has both domestic and external origins,' Herzog argued. 'And that economic crisis, and the fight against that economic crisis, has a social cost. The welfare and the standard of living of a good number of Mexicans has been dropping in the last years. But that is because of the crisis itself, not because of the measures that we have been taking.

'We have not simply followed instructions by the IMF. What happened was the Mexican diagnosis and the Mexican recipe and the Mexican solution were accepted by the International Monetary Fund. Of course when you are having a big deficit, like the one we were having, what you need to do is to cut expenditure and increase revenues, and that's what we did. But not because it was an imposition by anybody.'

For a time, it looked as if Mexico's approach of working with the banks and the IMF for a renegotiation of debt each year, acting the 'model debtor' as the American Government proudly called it, was working. Money was trickling back from abroad and the commercial banks seemed readier to make new loans, when the September earthquake forced the country to seek several billion in new loans just to cover the damage and the slide in prices in the winter of 1985–6 threatened to prevent it from ever reaching a position of being able to repay its debts. By that time it owed, including the new loans it had taken out in 1983–5, a staggering $96·6 billion, a sum that incurred interest payments alone of $10 billion. With oil at $28 per barrel, the price assumed in Mexico's rescheduling agreements with the banks, it could just about manage to cover the interest and repay some of the loans with a little new money from the banks. At $20 per barrel, it would need to take a savage dose of further retrenchment in its imports and its spending if it was to get through. At $10 per barrel, it was simply impossible. There was no way the country could ever afford to repay its bankers at that price. Only political intervention by the

American Government, or a decision on its part to walk away from its debts, could get it off the hook.

A Global Problem

Mexico, although by far the biggest of the oil borrowers, was not the only one. The years of plenty had seen all countries boost their spending and their consumption. Nigeria, with an economy three-quarters dependent on oil exports, had built up debts of over $10 billion and was forced in early 1986 to declare that it would have to limit repayment to 30 per cent of its foreign currency earnings. Venezuela and Indonesia, with similar debts, had cushioned themselves with higher reserves. But even they, at $10–15 per barrel, were talking of imminent disaster. Egypt was, in many ways, even worse off. The debt problem had become a political as well as a financial issue in the world. And it was a problem that could not be divorced from the price of oil.

In the days of high oil prices, as much as $100 billion a year had been transferred from the pockets of the consumers of the industrial world to the pockets of the producers. Some of those needed the money for themselves. Most of the biggest exporters deposited their oil revenues in the Western banks, who lent it back, at least in part, to the developing nations. 'Can you cope?' the then US Treasury Secretary John Connally demanded of Walter Wriston, the chairman of Citicorp, the largest bank in the world and one of Mexico's biggest creditors, as the first shock waves of the doubling of oil prices in 1973 hit the financial system. 'Yes,' came the confident reply. And the banks did cope. But inevitably the distribution of the wealth now deposited with them was uneven. Substantial deposits induced them to lend highly to sovereign nations on the principle, enunciated by Walter Wriston at the time, that 'nations never go broke'. In picking the countries to lend to, the banks gave special precedence to those with the resource whose price had risen so sharply – oil. Of the top two dozen borrowers, half were oil producers. Of the fifteen countries with debts of $10 billion or more in 1985, six were major oil exporters: Algeria, Indonesia, Malaysia, Mexico, Nigeria and Venezuela. For them the pain was doubly hard. Oil had fuelled ambitious spending plans. It had also induced high borrowings. When the prices fell, it hurt on all sides.

'You can divide the countries of the world into two types,' President Portillo was fond of saying to visiting journalists, politicians and businessmen, 'the ones that have oil and the ones that do not have oil. We have oil.' The epitaph to the follies to which such a philosophy gave rise, was voiced by one of the country's leading bankers. 'The benefits of oil,' he declared in 1986, 'were shared by the few. The miseries of debt are being shared by the many.'[5]

7

A Place in the Sun

In 1972 Sheikh Yamani came to Britain to address a major conference celebrating the establishment of the North Sea as a major oil province. The North Sea was no threat to OPEC, he declared.

> Such prejudice would have been probable in the fifties and sixties when supply exceeded demand in the oil market. However, now in the seventies, at a time when the market is on the verge of belonging to the seller, and the eighties not being too far off when the world might be subjected to a dangerous confrontation with the energy shortage, the situation is rather different. I now not only feel an absence of competition between Saudi Arabia and the North Sea but on the contrary, I feel pleased with discoveries of this kind which reduce the fear of the instability of supply and, perhaps, make the solution of the energy problem easier, even if only within narrow limits.[1]

Thirteen years later, in stark contrast, Sheikh Yamani emerged from a meeting of OPEC in December 1985 to announce that the North Sea was crucial to any price stability in the market. Unless Britain and Norway acted in accord to reduce output, the other exporters would be unable to stop oil flooding the market and the price collapsing.

What had happened to change the role of the North Sea from that of a marginal player on the world's oil stage to one of its leading actors? One answer was the oil market itself. The energy crisis that Sheikh Yamani foresaw in the eighties came instead in the seventies and brought with it its own response in surplus supplies and falling prices. North Sea oil, which had seemed small against a fast-rising world market, seemed much bigger against a falling one. In that sense, the world did return to the 'prejudice' of the fifties and sixties rather than Yamani's new world of the seventies. But in a deeper sense, the response of Britain and Norway to the value and to the role of their oil had also changed fundamentally.

Dividing up the North Sea

In 1964–5, when the countries of northwest Europe negotiated the division of the North Sea along the lines laid down by the recent Geneva Convention covering the Continental Shelf around their coasts, it was certainly hoped that oil and gas might be found. In northern Holland, Shell and Esso, operating in partnership in their exploration of northwest Europe, had discovered the huge Groningen gasfield, a field large enough to enable Holland, Belgium and much of West Germany, France and even Italy to convert from manufactured gas to natural gas. In England small pools of oil as well as gas had long been discovered and developed in the Midlands. Lord Cowdray was a firm believer in his later years that Britain would be the place of the oil future.

Prices of oil at the time were falling, however, and the technology of drilling for and developing offshore fields had only just begun in the Gulf of Mexico and on Lake Maracaibo in Venezuela. The main impetus was to secure the interest of the oil companies in the area and get exploration under way. During the mid-sixties the various countries surrounding the North Sea settled their agreements dividing up their offshore areas. The general principle followed was to define a 'median line' midway between their coasts and then to issue the acreage in 'blocks'. A few countries, like Denmark and Ireland, preferred to hand out their licence to a single consortium of companies on the basis that this not only ensured a partnership of local interest and foreign expertise but also, given the lack of knowledge and the unappealing nature of the area, proved the best way to attract exploration. Norway, the UK and Holland preferred to divide up their territories into blocks of 100 to 140 square miles each and award these to a variety of companies to encourage competitiveness among outside interests.

Strictly speaking, Britain might have had a case for claiming most of the Norwegian offshore area, including the whole area northeast of the Shetlands in which nearly all of the major oil discoveries have been made, since under the Geneva Convention, countries could claim the Continental Shelf around their coast up to a 'depth of 200 metres [657 feet]'. Around the southwestern coast of Norway, off Bergen and Stavanger, lies the Norwegian Trench, at twice that depth. 'A card thrown away by a government hell-bent on giving everything away in its rush to explore', was one of the accusations made against the Government of the time by later critics, convinced that Norway's major oilfields might all have been British.

It was most unlikely that Britain would ever have got away with it, even if it had tried that tack. Norway would certainly have refused to have come to any agreement with Britain on that basis and its claim that the Norwegian Trench represented part of its Continental Shelf and not the limit of it would certainly have been upheld by any international appeal. Nor was aggressive competition between neighbouring states the mood of the day. The North Sea countries, increasingly dependent on imported Middle East oil for their energy supplies as all of them saw the accelerating decline of their home coal industries, wanted to move ahead. Current geological theory pointed to the offshore as essentially an extension of onshore geology. On that basis, the interest lay in the area between Holland and the features of the Groningen gasfield and the English Midlands. The seas off Scotland and Norway seemed far away from possible oil and gas sources, the granite rocks on either coast highly unpromising and the water depths far greater than anything with which contemporary production technology could cope.

The First Discoveries of Gas and Oil

Indeed, the first phase of discoveries all did occur in the southern North Sea. It was gas not oil which was found and not in the extension of the Groningen features but in the coal measures of ancient forests. Oil, in contrast, is mostly found in marine sediments of prehistoric days. The gas discoveries were important enough in their own right. They enabled Britain to convert its entire gas industry to natural gas over the period 1968–73 and also helped develop a new offshore exploration in northwest Europe which was at the forefront of the technology of its day. What they did not reveal was any indication of oil. The understanding of Continental Shelf geology and the movements of the earth's crust (plate tectonics) was still too new. The belief that offshore geology might differ significantly from that onshore and that oil might have been generated during the separation process of the continents was only just beginning to gain credence. Sir Eric Drake, chairman of BP, was retailing the opinion of one of his senior geologists and the accepted wisdom of the industry as a whole when he said as late as April 1970 that he didn't rate highly the chances of finding oil in commercial quantities in the North Sea. 'There won't be a major field there,' he was reported as saying.[2]

Only a few months before Sir Eric's remark, in December 1969 after several years of fruitless and expensive effort, Phillips Petroleum had established what turned out to be a major discovery off Norway with the Ekofisk Field. As so often in the history of the oil industry, the group was thinking of giving up when its drilling at last hit oil in nearly 122 metres (400 feet) of water. And within six months of Sir Eric's statement BP itself had discovered the Forties Field, one of the largest oilfields to be found anywhere outside the Middle East, let alone on the doorstep of one of the world's biggest consuming areas. Within two years of that find, a series of discoveries by Shell-Esso, Armand Hammer's Occidental group, BP, Phillips, Mobil and others had established the North Sea as a major commercial oil province, one which might, thought the experts, contain anything between 20–50 billion barrels of recoverable oil and produce between 2 to 5 million barrels per day, enough at least to promise Britain self-sufficiency in oil and to make Norway a major oil exporter.

The discovery of oil altered the entire political environment of the

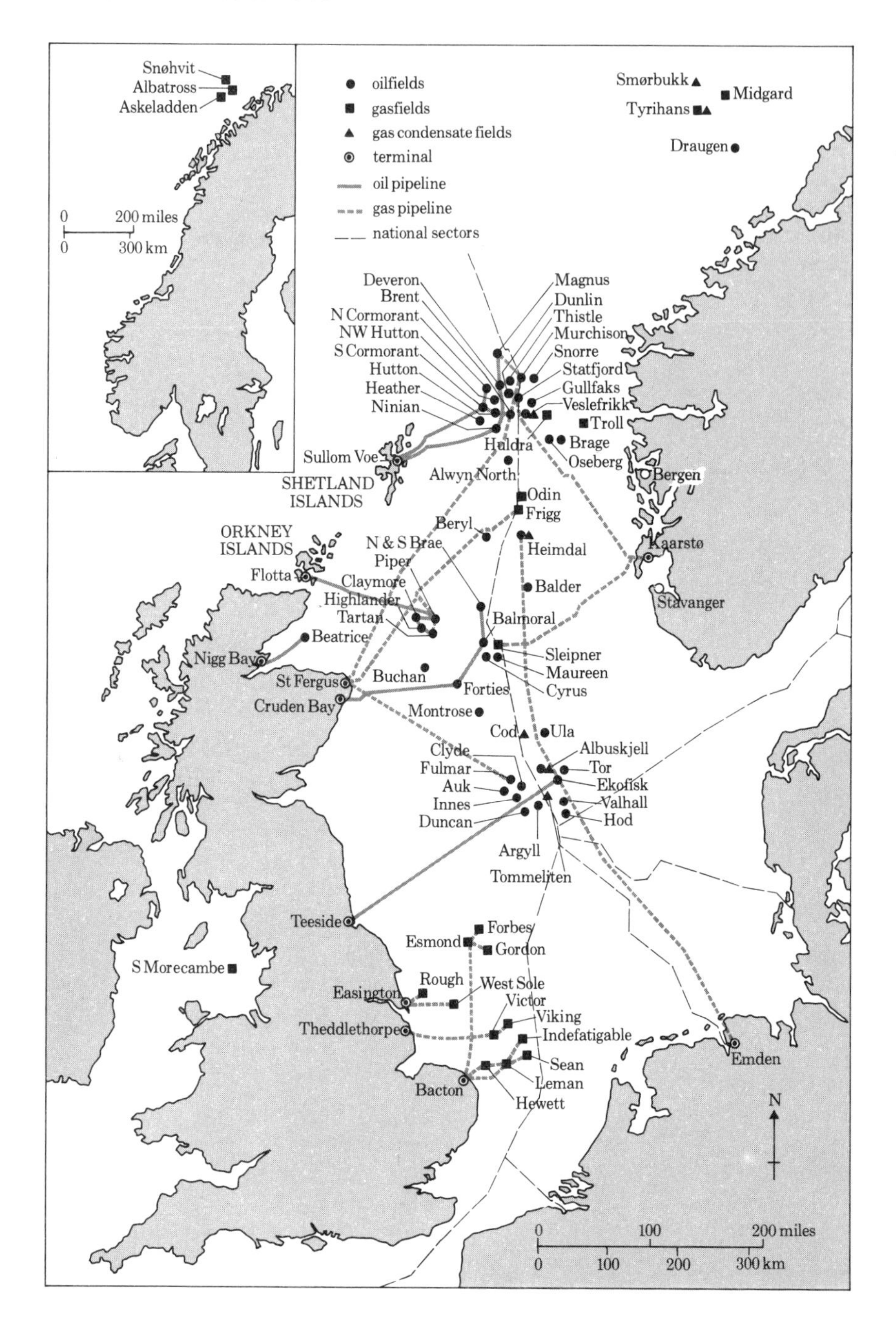

The main oil- and gasfields in the North Sea. Exploration started in the 1960s with the gas discoveries off the British coast, moved to the major oil finds northeast of the Shetlands in the 1970s and is now concentrated to the far north of Norway and west of Shetland.

North Sea countries. As long as it was only gas that was found, the question was one of how best to develop the resource and where to find the money to build the pipelines and the processing plants to take the fuel to market. Oil was quite different. It promised far more fundamental changes in the economies of the oil-rich countries, a far greater impact on the local communities nearest the finds and, of course, far greater potential rewards for the companies themselves.

The Norwegian Response

The approaches of Norway and Britain soon began to diverge. Norway, with a population of only 4 million, a strong sense of local community bred of the isolation of fjords and fishing ports and an inherent social concern in its politics, saw the prospect of major oil developments as something which was in part threatening to the Norwegian way of life and certainly needed to be controlled. Whereas it had followed Britain's example of relatively liberal terms in its original licensing, it now used the opportunity provided by Phillips' application to pipe the oil from Ekofisk to the UK in order to avoid piping it across the Norwegian Trench to Norway, to insist on greater state of participation in oil developments. A state oil company, Statoil, was established to keep the Government's share in the Ekofisk pipeline and to participate in future offshore licences and onshore oil and petrochemical investment. Tax terms were introduced for the industry. Finally, and most importantly, the Government undertook a thorough review of the whole question of oil's role in the country's economy and the speed with which oil development should proceed. In a series of reports to its parliament, the Storting, culminating in the pioneering Finance Ministry Report, No. 25 and the Industry Department Report, No. 30, during the 1973–4 sessions, the Norwegian Government laid down a policy of 'making haste slowly'. The granting of new licences, the rate at which new fields were to be brought onstream and the level of oil and gas exports were all to be controlled at a pace suitable to the absorption ability of both the economy and the local communities.

This policy and the tactic of using export permits as a means of tightening terms and introducing state participation, earned the Norwegians the reputation of being, in the words of the Texan oilmen, 'the blue-eyed Arabs of the North'. It was a reputation which the Norwegians half resented and were half flattered by. The oil companies hated it, in part because they realized the demands made on them were changing the rules in mid-game and in part because they saw in Norway's policy of control and state participation a development that was bound to encourage the Arabs to move along the same route. Somehow, they felt, an industrial Western national should behave differently. The Norwegian Government held firm to its policy for precisely the opposite reason. They regarded their problems in coping with a sudden explosion of oil wealth and investment as peculiar to Norway, rather than as part of an international development. If the North Sea were not controlled it would prove both unacceptable to the economy and dangerous to the community.

The British Response

Britain, at first, saw things quite differently. With a huge and complex economy, a population of over 50 million and a massive oil-import bill, the Government's original response to its oil discoveries was to accelerate the pace of activity, not slow it down. In 1971 the Conservative Government under Edward Heath offered an immense package of new licences covering most of its unallocated acreage of the North Sea. Its conscience pricked by a growing number of critics from both left and right who argued that the financial terms offered were grossly over generous to the companies, the Government also chose to combine this round with an experimental auction of fifteen blocks. In an atmosphere of hushed anticipation the companies' sealed bids for the blocks on offer were opened. Some of the blocks, chosen because they were representative of the various areas in the North Sea rather than because they were considered the best, received only small bids. But one, the so-called 'golden block' northeast of the Shetlands, was bought by Shell-Esso for a total, staggering at the time, of £21 million.

The auction only added to the political furore. The North Sea had been a 'give-away' by the British Government. Prodded by critics such as Lord Balogh, the Public Accounts Committee started an investigation in late 1972 and in the spring of the following year it produced a devastating indictment of British policy, pointing to huge tax credits built up in the Middle East which could be set against the North Sea, the lack of state participation and the failure to prepare special fiscal terms for oil. A fierce political debate on the nation's share in its oil wealth was already under way when the energy crisis of 1973–4 took the issue on to a new plane. Oil companies, making healthy enough profits before the crisis, now stood to make a veritable fortune. In 1974, after a prolonged and bitter miners' strike had caused Edward Heath to put the country on a three-day week, his government fell at the elections and a new Labour administration under Harold Wilson was elected, committed to 'nationalize the North Sea'.

After considerable debate within the Labour Cabinet, furious objections by the oil companies, backed by the American Government, and the threat of some groups to abandon the area entirely, the commitment to nationalize was reinterpreted by Lord Lever as signifying 51 per cent participation in all oil production rather than 51 per cent control of the equity of the North Sea groups. The oil companies were to sell just over half their North Sea oil to the Government who would buy it at prevailing market prices. The oil companies wouldn't lose out financially and the Government would have control over where the oil went, if not how much was produced, at little cost to itself. It was the true magician's sleight of hand, which infuriated the Labour party left, satisfied the oil companies and kept faith with the letter if not the spirit of the Government's election promise.

A state oil company, the British National Oil Corporation (BNOC), was established in January 1976 to participate in new rounds of licences and to take over the existing North Sea holdings of the National Coal Board and Burmah Oil, which had just been rescued by the Government from going into receivership. The Department of Energy, under Energy Secretary Eric Varley and his successor Tony Benn, further moved through the Petroleum and Submarine Pipelines Act of 1975 to enforce state participation in all new licences and to take more direct control both over the rate of depletion in the North Sea fields and the pace of new investment. Now it was Britain's turn to follow the example of Norway.

The Oil Companies and Nationalization

The oil companies were not pleased by the change. Two, recalls Tony Benn, who was Energy Secretary from 1975 to 1979, stood out: 'Amoco was one. They simply wouldn't go along with the participation agreement. So I warned them and I said, "This is the policy and you've got to go along with it." And when new acreage was announced, they didn't get any.

'Within a very short period they changed their management in Britain and came back and said, "There's been a terrible mistake." And they went along with the policy and were considered for the next round.

'In the case of Exxon the world president came to see me. He sat down and said, "I'm not prepared to discuss your policy with you."

'I said, "Why not?"

'And he said, "Well, the political philosophy of this company is different from the political philosophy of the Government!"

'I could see the faces of all his advisers absolutely falling in dismay. And I said, "Well, thank you very much indeed for telling me that you as Exxon regard yourself as entitled to dictate a policy to an elected government. As far as I am concerned that's the end of the matter." And I wouldn't discuss oil with him for the rest of the meeting. He invited me to meet him again a year later, which I did. I declined to discuss oil with him.

'In the end Exxon came round because these oil companies are very vulnerable. Nobody should be obsessed by the strength of the Seven Sisters in the face of a determined government. A determined government has got the power, if you have oil as we had, to decline to give them acreage. And secondly you have the power to withhold the market, because remember, Britain was an important oil territory long before we discovered oil.'

For the British, perhaps more than the Norwegians, whose attitude towards oil was coloured by the need to control the whole development as much as to ensure a state share in it, the issue became very much a battle between the national state and the foreign oil companies. Tony Benn became something of a bugbear to the companies partly because he so constantly emphasized this point. The 1973–4 energy shock clearly altered the balance of advantage in such a struggle from the company, courted in the days of oil surplus, to the Government, courted in the days of shortage.

'Scotland's Oil'

The 1973–4 energy crisis, however, didn't just alter the balance of negotiating power between governments and companies. It also altered the politics of oil within the North Sea countries themselves. The fact that most of the discoveries had been made off the coasts of Scotland gave enormous impetus to the nationalist movement there, then growing in strength and threatening to unseat the traditional Labour and Conservative representation in the country. The Scottish Nationalist Party seized on the slogan 'It's Scotland's Oil', and for a time in the post-1973 era, it seemed destined to become a major force in British politics.

' "Scotland's Oil" was one of the best slogans that's ever been devised in modern politics,' argues Gordon Wilson, the present leader of the party. 'We made the case right at the start that Scotland would only benefit from the oil if we did have our own government . . . we knew right at the outset as part of the United Kingdom that unless we did assert a national claim to a share in the oil, we'd be left with peanuts.

'We did extremely well in the 1974 General Election when we got not only eleven members of parliament but 30 per cent of the Scottish vote. It was the biggest vote any third party has ever achieved within the United Kingdom.' It was a vote strong enough to convince a number in the Labour Cabinet, including the future Prime Minister, James Callaghan, that both Scotland and Wales would have to be given a measure of devolution if the threat to the unity of the UK was to be avoided. Others, including Tony Benn, argued that the issue was a complete irrelevance to the problems of poverty and class. 'I think the illusion,' he maintains 'that if a man in a kilt owns all your land or owns your oil, that's going to benefit the working class in Aberdeen or Glasgow or Edinburgh is a complete illusion.' And he readily accuses the oil companies of having deliberately stirred the waters of Scottish nationalism 'because they knew that if Scotland broke away it would be easier to control and similarly the oil companies were very keen on the Common Market because they always tried to use the Treaty of Rome to break the policy that we had for bringing them into line with our interests.'

The Shetlands and Sullom Voe

The added problem for both the London Government and the Scottish Nationalists was that a great part of the oil discovered lay not simply off the coast of Scotland, but more particularly off the Shetlands. The Shetlands, a group of islands with a population of just over 17,500 at the time of the oil discoveries, had always regarded themselves as much a part of Scandinavia as of Scotland. They were in no mood now to let the Scottish mainland seize the benefits, or to lose out themselves. When the oil companies first approached the local council with ambitious plans to build pipelines from the major fields such as Brent in the East Shetlands Basin, the islanders were initially caught without much knowledge of what was going on or any clearly defined policy to deal with it. With the help of an able and determined County Clerk and general manager, Ian Clark, a mainlander from Hamilton, and local councillors including Edward Thomason, the Shetland Islands Council sought outside advice and prepared itself for negotiations with the oil companies, and even more important, a struggle with the central Government to ensure that the council gained the power to control the developments. Like the Norwegians of Stavanger and Bergen, Shetlanders might be prepared to welcome the advent of the oil industry for the jobs it would provide and the money it would bring. They were, however, extremely worried that the size of the facilities being sought by the oil companies and the number of outsiders that would be brought in to construct a terminal for the new oilfields could swamp the islands with disastrous results.

'It should always be remembered,' emphasizes Ian Clark, 'that at no point, right up until oil came to Shetland, did the Shetland Islands Council encourage oil or suggest to the people of Shetland that they would be better off with oil coming. The council's viewpoint was that Shetland would be better without oil arriving but that, if the island was needed for oil, central Government would make sure that it was used for oil. So we shouldn't embark on a great long debate on, "Should we have it or should we not have it?" We should embark on a great debate about, "If it is going to happen, what are the minimum terms that we should be demanding before we allow it to happen?"'

It was this determination to make the best of what was an unavoidable, but in some ways unpleasant, job that made Ian Clark such a formidable negotiator on behalf of the islanders. Only 29 years old at the time, with no experience of big business, let alone the oil companies, he none the less had the advantage, on his own account, of the conviction that what he was doing was right. 'I was never in any doubt', he accepts, 'but that these gentlemen were in a completely different league from myself. But I was never in any doubt as to the worth of what we were trying to do in Shetland and they knew that, even if they were better businessmen and better intellectually than I was. The one thing I knew better than they did was, first of all, what the Shetlanders wanted, and secondly what I thought the Shetlanders needed. And they showed themselves so singularly unable to apply their mind to that particular point. That was a great benefit to me because, in terms of public perception, they were constantly wrong-footing themselves. Shetland won the media battle primarily because the industry representatives were so poor at understanding what an isolated community thought.'

He recalls one incident very clearly. 'The convenor of the day and myself were invited to Shell Centre for further discussions on what they regarded as our unrealistic demands. They tried to move us, as they had done many times before, and ended the session by saying that they regretted to tell us that our demands were so unreasonable that they would have to go elsewhere. I responded for the council and said, first of all, that I wanted to congratulate them on their technical ability to go elsewhere. Secondly, I wanted to assure them that they had the support of the council representatives in going elsewhere and that, if we could be helpful in introducing them to our colleagues elsewhere in the United Kingdom, then they had only to tell us and we would willingly do that. I ended by telling them that, while our discussions throughout the months had tended to be acrimonious and, from time to time, less than absolutely pleasant, I was delighted that our discussions were ending on this extremely happy note because they were sending two happy men back to Shetland. They were sending us back with a message that by far the majority of Shetlanders wanted to hear, that the oil industry was not coming to Shetland. At that point we thanked them for their courtesy in dealing with us over the months and stood up to go. When we reached the door, we were unceremoniously put back into our seats and told that the oil industry had to come to Shetland to deal, so, "Let's get it all sorted out."'

It took the council four years to achieve both the powers it wanted from London and the terms it sought from the oil companies. The first was finally obtained by means of a parliamentary bill, the Zetland County Council Act of 1974, which gave the Shetland Islands Council wide-ranging powers to acquire land, construct port facilities and set up a charitable fund to invest the oil revenues for the future. It had to be fought against the opposition of local commercial interests, which wanted to develop facilities privately, as well as oil companies and supply groups. The agreements with the oil companies proved even more difficult to negotiate. The council had picked a site suitable for the proposed oil terminal at Sullom Voe, a name derived from the old

Norse *sol-heimr* meaning 'a place in the sun'. It now proposed to purchase the land, lease it to the companies and develop harbour facilities under strict conditions and tough financial terms.

The council's aim was, like that of the Norwegians, to minimize the immediate impact on the local community and to ensure a sufficient return to enable them to build up other businesses for the future when the oil flows might cease. The oil companies, which at first acted independently, tended to see the council's demands as just another example of government extortion and they appealed against them to both the Energy Department and the Government. The Shetlanders, however, were in no mood to be bullied and the Government had been sufficiently unnerved by the threat of Scottish nationalism in the 1974 elections not to upset local feelings too far.

Talks with the companies continued over the period 1974–8. Indeed they were still incomplete when the first oil was pumped into Sullom Voe in November 1978 to be shipped out in the Shell tanker, *Donovania*. A licence, issued to the oil companies to allow them to operate the terminal was only accepted subject to 'important qualifications'. By then the companies, after considerable bargaining, had agreed to pay the Shetland Islands Council 1p per tonne under a Disturbance Agreement intended to compensate the council for the extra costs involved in coping with the oil influx and the community unheaval it would cause. In addition the companies agreed to lease the 1012 acres of land from the council at Sullom Voe and to pay it for constructing harbour facilities. A joint company, the Sullom Voe Association, was established to operate the terminal by the community and the companies.

The agreements were hailed at the time not only as a victory for the Shetlanders but also as a path-breaking effort by a small and rural island community to shape and control the impact of a huge and sophisticated new industry invading its shores. The financial terms and clauses, said Sir Peter Baxendall of Shell, celebrating the signing of the Sullom Voe Association Agreement in April 1975, went 'beyond what was normal commercial practice'. 'A unique partnership' had been formed under which the council would have 'a controlling voice in all those affairs which affected the quality of local life'.

'There is absolutely no doubt,' believes Ian Clark, who in many ways could be regarded as the architect of the deals, 'that what Shetland achieved was, in the eyes of all knowledgeable onlookers at that time, unbelievably unique. Not only did it bring about a unique state of affairs in Shetland but it changed very radically the relationship between the oil industry and communities elsewhere in the world. You have, over the years, had people coming from as far afield as New Zealand to look at what happened in Shetland and to try and understand how it happened.'

At the time of writing the council is about to take the oil companies to court in an atmosphere of bitter recrimination. Oil revenues had dropped far below expectations. The council was deeply in debt and its much-vaunted fund had failed to produce the results many of the islanders had expected. The oil companies were accused of having 'conned' the Shetlanders. For their part, the companies accused the council of over optimism, over spending and over indulgence. 'The place in the sun' had begun to seem an illusion and a foolish one at that.

A major part of the problem was the changed conditions of the oil market. When the Shetland islanders had first negotiated with the companies, the whole North Sea had become overtaken by an atmosphere of mounting bonanza. Shell's £21 million bid for the 'golden block' had signalled a string of major oil finds in the East Shetlands Basin. Discovery succeeded discovery on both sides of the median line between the UK and the Norwegian sectors at Brent, Dunlin, Cormorant, Ninian, Hutton, Thistle, Murchison and Statfjord. As the oil companies flew in to talk to the Shetland Islands Council about landing possibilities, as the price of oil doubled and then redoubled and as the number of finds mounted, any word of caution by the more conservative companies tended to be dismissed as typical oil company secrecy and negotiating ploys. When serious discussions about the proposed terminal finally got under way, there was a real prospect that the finds would justify the construction of three pipelines with a capacity of 3–4 million barrels per day in total, or 150–200 million tonnes per year which would bring in an annual income of £4 million.

In the event, the terminal opened more than a year behind schedule. Although its initial capacity of 0·8 million barrels per day was uprated to 1·4 million barrels, the third pipeline was never built and the council failed to gain either the income from the Disturbance Agreement or the Ports and Harbours dues that it had expected. At the same time its revenue from the rates on Sullom Voe, which account for nearly two-thirds of its total income, suffered a savage cut when the terminal became subject to 50 per cent and then 40 per cent industrial derating, reducing its potential valuation from £60 million to £18 million in 1984. The council's total income from 1975 until the end of the 1984–5 financial year totalled £174·753 million, of which £110 million came from rates and the remainder from the Ports and Harbours and disturbance payments. 'By the year 2000,' bleakly admits Edward Thomason, 'Shetland will have received £200 million less than we anticipated.'[3]

In the meantime the council simply went on spending. Debts totalling £140 million had been piled up, mainly, the council argues, to build houses, roads and other facilities to cope with the influx of oil companies and contractors. Domestic rates were kept low. In addition nearly £35 million was spent from the council's reserve fund and its special charitable trust on industrial and commercial projects such as new fishing vessels and fish factories. A further £22 million had gone on social work, leisure projects, housing and recreation. As much in desperation as in disappointment, the council finally went to court to fill the gap through a claim for rent on its Sullom Voe land of £215 million in back payments and £90 million a year from 1985. The lease arrangement had never been finally agreed between the parties and in the end the oil companies had simply gone ahead under a licence that left this part of the deal open. The council now claimed that the lease should cover the full value of all the facilities built on the land. The oil companies replied firmly that the understanding had always been that the rent should be based on the value of the land itself, a calculation that would give the council only £300,000 a year instead of nearly £100 million. The atmosphere was hardly helped by the fact that the 'Voe House Letter' of 1976, setting out the terms, was kept secret. With barely concealed feelings of disillusionment on both sides, the lawyers were briefed and the cases prepared for hearing in 1986.

'I feel no guilt at all,' says Basil Butler, a BP director closely involved with the negotiations with the Shetlanders during the seventies, 'if there's a feeling that mistrust has grown up between the Shetland Islands Council and the industry. In fact from the very beginning I think there's been a feeling of mistrust, which is perhaps understandable in view of the size of the operation that went on there and the large companies coming in to this fairly small community. . . . Undoubtedly the industry made some mistakes. And so, I would contend, did the Shetland Islands Council in the early days, which has led to a feeling that neither side has been as easy with the other side as one would have hoped. And this is a great pity.'

That is not the way that most Shetland islanders saw it at the beginning, or see it now. 'The first planning application for a construction camp,' recalls Edward Thomason, 'was the one at Mossbank at Firth. It was for a camp for 1200 men for five years. Then the plans for the terminal again expanded and the industry applied for a further work camp of 1800 men for eight years. Then two construction ships had to be brought in, each capable of accommodating 500 men. So the size of the terminal grew and the workforce grew from 1200 men to 4000. It was a time of uncertainty and Shell gave what they stated was their best forecast. We believe that they were better equipped to make such a judgment than we. And we made the agreements on the due date of every activity that involved planning permission. We cut corners in order to help them.'

Far from being greedy, Mr Thomason argues that the council's demands reflected no more than what was needed to help what was a 'small rural community' to cope. 'We had roads,' he says, 'that everybody said were inadequate – very pleasant to drive over because they are so bad nobody could hurry. We've had to build schools for a new population. We've had to build houses for a new population. We've had to build roads. All this was compressed into a very short period of time. We have had to build in ten years what this community might normally have built in twenty-five to thirty years.'

The problem for the Shetlands has been, of course, not simply whether they gained as good an agreement from the oil companies as they had hoped, but whether they have been able to use the wealth of oil to secure their future. 'After the oil is over,' Thomason points out, 'in twenty or twenty-five years, we will have a new population of some 6000 additional people here in Shetland. And many of them will have, of course, to stay in Shetland. Once upon a time people could leave. But where do you go to now, at least in Western Europe, where you can get a job? So we are left with a social problem of how to deal with the people. Their parents may not be Shetlanders. The bairns are Shetlanders. They're born here subject to the same disciplines, the same background. Now we've given a good service to the oil industry. We've provided them with a joint user base at the terminal which they might never have been able to achieve themselves. They've done well by the Government and we see no reason, in return for these services, why we shouldn't arm ourselves for the coming years.'

Twenty years ago, Thomason points out, 'young Shetlanders had no jobs here, so they went whaling – two to three hundred of them. They went to Antarctica each September and came back the following April–May. The same number were in the Merchant Service. The British Merchant Service is now more or less defunct. So is whaling. But there are no jobs in Shetland for the young.'

To that extent, the coming of oil could never have been a simple issue for the Shetlanders. On the one hand, it threatened the community with an influx of outsiders that would have swamped its own local inhabitants and put at risk the traditions and balance of the island. At one point the council was forced to put up posters in railway stations throughout Britain to advise against people coming to Shetland because the islands had neither the housing nor the jobs for them. On the other hand, oil has brought the Shetlands jobs, money and facilities that they would never otherwise have gained. By setting up a charitable trust the council attempted to hold back some of the impact of oil wealth and invest it for the future. Their expenditure on sending sports teams abroad and building expensive sports facilities at home has aroused strong criticism from some who argue that much of the wealth has been squandered on idle spending and high subsidies to keep the rates down. Yet, given the presence of the oil revenue and the expectations of ever larger volumes, it is hard to see how any democratically elected council could have held it all back for the future. As it was the charitable trust was established with relatively tight controls on spending and, by sensible investment, had nearly doubled its funds by early 1985 from £27 million to £51·5 million.

Even though it proved considerably less than expected, an annual oil income of over £50 million among a population of 22,000 would have seemed reasonable enough to most people in Scotland. What couldn't be easily coped with was the sudden contraction in the industry's exploration and development budgets and the prospect of a much earlier than predicted decline in output through the pipelines to Sullom Voe which came with the oil price fall in 1986. Rightly or wrongly, the islanders had grown used to the idea that their spending, evenly controlled, would last a generation, not less than a decade. As Edward Thomason comments: 'It may well be that we have now to try to temper the diet of oil money.'

Norway's National Economic Emergency

Lying halfway between the UK mainland and Norway, in many ways the Shetlands symbolized the dilemmas of both. In the first rush of oil discovery in the *anni mirabilis* of 1970–4, it was as if nothing could go wrong. The only question was how best to respond to it and protect the national interests. In the space of those four years, drilling in some of the most hostile conditions anywhere in the world, enough oil was proven to fulfil the predictions of even the most optimistic in 1970. The British could safely assume a production level of around 2–3 million barrels per day by the end of the seventies. The Norwegians were able to plan to restrain deliberately the production of oil and gas to a political ceiling of around 1·5 million barrels per day equivalent.

As it turned out, the pace of production was slower than planned. The costs of building and installing facilities in such deep waters in the half year when the weather was good enough to work offshore were far greater than first envisaged. Most of the early field programmes started one to two years behind schedule. But if production was pushed back a little, it still eventually reached the rates originally expected. By the time of the second oil shock of 1979–80, Britain had achieved net self-sufficiency in oil with an output of 1·6 million barrels per day and by 1984 she had become a net exporter of oil with a total output of 2·5 million barrels per day. By then Norway was producing over 700,000 barrels per day, four times its internal consumption. By then too, oil revenues had climbed to a point where they made up well over 10 per cent of the UK's total revenues and some 5 per cent of the country's total output. In Norway, over the same period, oil had come to represent well over a third of exports, 16 per cent of total public revenue and 17 per cent of total output.

When oil prices started to fall in the mid-eighties, both Norway and Britain were caught on the hop. Both countries had benefited enormously from the oil price rises but both now found themselves entangled in the general recession of the industrial world induced by the 1973–4 and 1979–80 price rises. Norway, which had been relatively slow to adjust to the industrial restructuring caused by the first oil shock in 1973–4, was forced into the unparalleled step of calling a national economic emergency in 1978.

'In the seventies,' recalls Oystein Noreng, Professor at the Oslo Institute of Business Administration and a former official with Statoil, 'a bit naively we thought we were on a very quick road to very easy and overwhelming prosperity. So we thought we could spend money to get away from the recession. And we overspent and we had to call in 1978 a national economic emergency. For the first time in peace in Norway we had to restrict the right of trade unions to negotiate wages. There was a mandatory prices', and wages', freeze, a kind of economic dictatorship. In 1978 the foreign debt, the public debt, represented about half of the Norwegian GNP. Much of that was sunk in oil. But money had also been spent on saving the shipping industry and also on keeping consumption high. Real wages in Norway rose by about 25 per cent between 1974 and 1977. That's a high increase for any three-year period.'

The reaction comprised both a package of draconian measures to halt inflation and a revision of the oil policy to allow rapider development and a more extensive issue of new licences in the highly prospective but extremely expensive waters off Norway's northern coasts. It also led to renewed efforts to formulate policies which would keep the oil-income flows from upsetting the balance of the Norwegian economy too much by keeping part of the money abroad and by accounting separately for oil income in the budgeting process. The aim was to avoid what had become known as the 'Dutch disease', after the Dutch experience of increasing social welfare payments and wages disproportionately in response to the dramatic earnings from the Groningen gasfield in the sixties and seventies.

Wasted Opportunities

Britain had had its economic crisis package earlier in 1974, in response to the coal strike of that year. A succeeding Labour Government then attempted, ultimately unsuccessfully, to restrain wages and public spending on a medium-term programme agreed between government, unions and industry. It was not until the second oil shock of 1979–80 that Britain received the first real boost of rising revenues from oil and substantially increased export earnings. By then a Conservative Government under Mrs Thatcher was set on a monetarist course. Rising oil earnings were allowed to raise radically the value of sterling on foreign exchanges, but government spending was kept tight and the money supply controlled almost as if there was no oil.

The combination of a high exchange rate and a squeezed domestic economy had a devastating effect on industry, accelerating the industrial shake-out occurring throughout Western economies and driving a large number of manufacturing companies to the wall. ICI went to the Government to warn it that it would be forced to get out of most of its export markets and close much of its UK plant if there was no change in policy. 'What we had with the advent of oil,' argues William Keegan, one of the most persistent critics of government policy,

> was the chance to cushion the balance of payments against any kind of shock from a burst of expansion. What we actually did in the end was to have a period of slower economic growth during the period 1979 to 1985 than in any five- or six-year period since before the Second World War. We completely neglected to use the opportunity of the oil.[4]

From the Government's point of view, however, it was oil that enabled it to pursue a tight fiscal course by sustaining its revenues, which would otherwise have collapsed with the contraction of industry and the accompanying rise in unemployment. Oil paid the unemployment benefits, claimed the critics. Not so, replied the Government, it helped the process of the long-needed adjustment of British industry. Far from being wasted, by allowing so much money to flow freely abroad during the years of high exchange rates, the country was building up investment income that would serve it well in the future.

When oil prices did start to dip in 1983, therefore, both Norway and the UK found themselves ambivalent in their responses. On the one hand they had no wish to appear part of a producer cartel to keep prices up against market pressures. On the other hand, they both wanted prices kept up, preferably by OPEC. Norway by this time had eased up its economic policy following the 1979–80 oil price increases and was enjoying relatively healthy rates of growth. Britain had come to depend on the £12 billion a year in government revenues it gained from oil to cushion itself against the full blast of recession.

In March 1983 OPEC deliberately chose London as the venue for its crucial meeting to decide on oil production cuts and prices. As Sheikh Yamani argued to British ministers and the cameras at the time, the British couldn't for ever have it both ways. Rising North Sea production at a time of falling demand was inducing fierce competition in the crude-oil markets with Nigeria and the Gulf. Britain was now the fourth largest oil producer in the world and both Britain and Norway were among the top ten oil exporters. Their interests should lie with OPEC, not with the oil importers.

It was an argument which both countries formally rejected. Both were part of the major consuming areas. Neither country was in a position to adjust current supply easily, since the production rates of each field were decided by the companies not government. Behind the scenes, however, both governments quietly used their influence through Statoil and the BNOC to ensure that neither led the way to oil price cutting. When OPEC finally reached agreement on production quotas in London and prices stabilized around new, but only slightly lower levels, the governments of Oslo and London breathed a sigh of relief. It didn't last long. Within a year a new round of price cutting was developing in world markets and this time it was the Norwegians, through an inadvertent price decision by Statoil, which set off further falls. This time the British Government took the extraordinary step of writing to all the oil companies asking them to act with restraint and hold prices.

The experience finally ended the British Government's appetite for intervention. The role of the BNOC as a 51 per cent purchaser of all oil in the North Sea was proving more and more costly as prices fell and it was coming under increasing parliamentary criticism. In the autumn of 1984 the Government announced that it was abandoning the role altogether and unwinding the participation agreements.

Sheikh Yamani's vigorous demand that Britain and Norway join this time in OPEC production restraint as the precondition of any effort to hold up prices during the winter and spring of 1985–6 found the previous roles of the two countries curiously reversed. Norway, which at the beginning had seemed to fear the impact of oil, now saw it as vital to its economic well-being. Its response to Yamani's call, while uncertain, was none the less sympathetic, all the more so after the Conservative Government fell in April 1986, to be replaced by a Labour Government which announced, as its first oil policy decision, a wish to co-operate with OPEC. The British on the other hand, who had started by greeting oil development as the greatest economic benefit since the Industrial Revolution, now felt that the general benefits to trade and to world interest rates of a fall in oil prices outweighed any disadvantageous effect on the country's balance of payments and income. In a series of public statements both the Chancellor of the Exchequer, Nigel Lawson, and the Prime Minister, Mrs Thatcher, made quite clear their refusal to join OPEC in holding prices.

'The interests of Britain are totally different from those of the members of the OPEC countries,' argues Alick Buchanan-Smith, the Minister for Oil at the Department of Energy. 'We're a consuming nation with a large population and with a huge industrial base apart from oil. . . . Don't forget that the purpose of OPEC has been a price cartel to maintain the price of oil. You talk to some of the major energy-consuming industries of Britain and you will find that that is something which they certainly do not see as in their interest.'

The Future

It is all in stark contrast to the mood a decade or even five years previously, when the two oil shocks had made Britain and Norway feel that theirs was the future, that they could ride the recessions better than their competitors and ensure their populations a long-term vision of prosperity. The oil has certainly proved to be there, along with equally exciting reserves of natural gas. For Norway, which has now made some potentially large oil and gas discoveries in the newer territories to the north, and has large amounts of unlicensed prospective acreage to put on offer, there is still room for growth and the prospect that, by the 1990s, it might be one of the most important net exporters of oil outside the Middle East. For Britain the future is complicated by the fact that, on present trends, its oil production will start to decline from 1987–8 onward. It will require the development of a host of new fields to prevent the country ceasing to be a net exporter by early in the 1990s.

The difficulty for both countries is that their North Sea future is based on high-cost investment which looks progressively less economic under low oil prices. Yet the experience of both countries is that high oil prices remain a mixed blessing for them as much as for their non-oil trading partners. For the general public as much as for government ministers and officials, economists and commentators there remains an uncertainty as to just how much benefit North Sea oil has brought them. 'We would have been better to have kept the bloody stuff in the ground,' exclaimed Michael Edwardes, former head of British Leyland, at a meeting of the Confederation of British Industry in 1982. He was loudly cheered. That didn't, of course, answer the political question of how, in a democracy and a free economy, you did keep the 'bloody stuff' in the ground once you'd found it.

8

The Global Gamble

In the mid-seventies, after the first oil shock, the oil companies started beating a steadily wider path. This took them as far as the doors of Beijing in China, where the Government, under its new liberalization programme, was talking of issuing its first ever round of oil licences. The prospect of drilling offshore China was big news to the industry. The country had an immense and virtually unexplored Continental Shelf. From the little that was known of it, much of the area looked highly prospective, with deep sedimentary layers and large structures in which oil might be trapped. What was more, China, a Communist nation that had closed its face to the outside world for more than a generation, had the largest population in the world. Once released on to the market, its 800 million people could form a customer base the like of which Western companies had not seen for a century or more.

The Chinese took their first steps towards opening up the country for oil cautiously. They invited companies to undertake seismic surveys of the area to test its geology and share the knowledge with them. Then, on the basis of these results (which were reasonably optimistic) they awarded large tracts to a series of chosen groups from several nations, including both some of the biggest companies in the industry and some of the smallest. The terms were strict. China made it very clear that it regarded all its natural resources as the property of the nation. They were prepared, however, to allow reasonable profits. Equally important to the companies was the fact that the Chinese officials declared themselves genuinely anxious to bring in foreign expertise and technology. Given the chance of getting in on the ground floor on the world's largest major untapped oil region, on the doorstep of its biggest potential market, most companies were all too anxious to apply.

After the drilling of some three dozen wells over the four years following 1980 when drilling had started, the mood changed. There were indications of oil and even a few small finds. But the South China Sea, the area first opened up, proved a let-down. 'It is most unlikely,' says Basil Butler, the BP director in charge of the area, 'that there will be any major fields discovered in the South China Sea, certainly in the shallower water areas. Nobody really has looked in the very deep water areas at the moment.' For oil exploration to be successful three things need to come together: first, the right conditions under which oil can be generated in ancient sediments; second, the right rocks through which it can move, or 'migrate', and third, the right trapping mechanism, or structure, in which it can be gathered in a reservoir. China had the last two. But, according to Butler, 'The biggest disappointment has really been the lack of good source rock to generate the oil to fill the structures . . . this is, of course, disappointing to us and disappointing to the industry as a whole because we all looked upon it as a very prospective area before we started.'

The story isn't over for China of course. The South China Sea is just one part of a vast onshore and offshore area containing a large number of potential oil provinces. Drilling has at least shown that oil does exist offshore, while oil has been produced onshore for decades now. Following the first drilling in the South China Sea, the Beijing Government has been talking to the companies since 1985 about concessions in both the Yellow Sea and further out in the South China Sea. The Chinese authorities started with the view that foreign companies would only be allowed offshore, partly because they felt that was where outside technology was most needed and partly because they wished to limit the internal impact of foreign investment. Now, however, they are suggesting that they will allow onshore drilling by the companies in the southern ten provinces, as well as easing the financial terms. On the horizon too, remains the possibility that they could open up the north and west of China where, according to Butler and other oil experts, the prospect of large finds could be very good.

That such high initial hopes should prove so unfulfilled, however, serves as a timely reminder of just how uncertain a business the oil search remains. It has had some remarkable successes where most of the experts had predicted little but wasted effort. Indeed, it could be said that the majority of the major oil finds of the industry's history – from Drake's well in Pennsylvania, Spindletop in 1901 and the East Texas field in the 1930s, to the finding of oil in Persia and, later, in the 1930s in Saudi Arabia and the discoveries in Libya and Mexico as well as in the North Sea and Alaska – have been made by determined individuals or companies pressing ahead against the firm consensus of the industry at the time. Standard Oil said that oil would never be found in the southwest of the United States. BP said it would never be found in commercial quantities in the North Sea. But then, even after the development of increasingly sophisticated exploration techniques, the rapid expansion of offshore technology and the dramatic improvement in seismic interpretation brought about by the use of computers, most geologists were surprised by the lack of success in the South China Sea. The 'golden block' for which Shell-Esso bid £21 million in the North Sea auction of 1971 proved to contain a far more complex and less profitable field than they had thought, although oil was found in huge fields all around it. As recently as 1986, exploration on Norway's 'golden block', failed to find the giant field which the oil companies and the Norwegian authorities had confidently stated would be there.

The results of initial drilling in China were frustrating in a wider sense too. Had China found substantial quantities of oil offshore, or even just one large field, it would have revolutionized the country's financial future. Foreign loans would have been much easier to obtain. The economy could have been expanded without the constraint of balance of payments pressures. The Government could have afforded to ease the controls on private car ownership and transport in the knowledge that it had the oil to feed them. The industrial take-off, which the reforming Government of Deng Xiaoping so urgently sought, would have been that much more possible with oil.

It would not necessarily have meant that the Chinese Government would have used the oil wisely. Mexico, Iran and, on a much smaller scale, the Shetland Islands, remain tragic examples of what can happen to developing countries which, for all the best of reasons as well as the worst of corruption, try to use sudden oil wealth to break out into rapid growth. Oil has brought economic happiness to neither a populous country that needs all the oil revenues it can get nor a desert kingdom that does not. In China's case, however, the road remains infinitely more difficult without oil. Even at the much lower prices currently prevailing, oil would have dramatically improved China's prospects just when it most needed it. As it was, without oil, in 1985 the Government was forced by a suddenly accelerating balance of payments deficit to pull back drastically on its investment and expansion plans, particularly those requiring foreign investment. Oil remains a key which can unlock the doors to foreign currency, foreign expertise and foreign investment.

It is this dual aspect of oil – the extraordinary risks and rewards of exploration coupled with the wider economic rewards and risks of production – that makes it so unique a commodity. When the first prospectors went in search of it in the United States, they looked on it very much as gold. Indeed many of them went straight from the goldfields of California and the Yukon to the oilfields of Pennsylvania, Texas and Oklahoma. If anything, oil was even easier to prospect for than gold or diamonds. Once you'd found it, it came up of its own accord at relatively little expense. The first oil cost Colonel Drake around 10 cents a barrel to produce and he could make a handsome profit by selling it for $20 as a substitute for kerosene manufactured expensively from coal or shale.

You could not, however, take oil from your well and sell it at the local assay office as you could gold. Oil is worthless until it is refined into products and sold into markets. Unlike gold or diamonds, it has no intrinsic value. To develop oil you need pipes, refining capacity and distribution facilities. Hence it was not the prospectors, like Colonel Drake, Patillo Higgins, Dad Joiner and Frank Holmes, who made their fortunes but the men with the money and the organization – the Mellons, the Hunts and Rockefeller. Rockefeller added something else, the understanding that oil above all commodities required integration if the resource was to be developed along with the market. He took it from the refinery and worked backwards into pipelining and then production and forward into distribution and marketing. Others did it from different ends. Henri Deterding started with production and used Samuel to develop the marketing side. Samuel did things the other way around. Many of the daughter companies of the Standard Trust, such as the Standard Oil Company of Ohio and Standard Oil of Indiana went from the market back to the source. Several of the European companies did likewise.

The constant problem of oil for the companies has been that, however hard they have tried, they have never been able to control exploration. As both Rockefeller and his great rival Deterding found to their cost, new discoveries constantly had a nasty habit of creating new sources of cheap oil and new rivals. Spindletop opened up the American southwest just as Rockefeller seemed impregnable in his dominance of the northeast. Iran made the fortunes of BP and Mexico the fortunes of Lord Cowdray when the power of Shell seemed all pervasive. At the time BP joined with Shell and Esso to take control of the Middle East, Saudi Arabia was opened up to produce a competitor in the Caltex partnership of Texaco and Standard Oil of California. When the Seven Sisters – Shell, Esso, Mobil, Gulf, Socal, Texaco and BP – seemed immovable in their control of Middle East supplies in the late fifties and early sixties, Libya, Indonesia, Alaska and finally the North Sea were developed and brought in other international names from the United States: Occidental and H. L. Hunt in Libya, Arco in Alaska and Phillips in the North Sea.

Even though the Seven Sisters were able to dominate the post-war oil market – and they were able to sit astride it to a remarkable degree given the growth in demand and the sheer scale of the industry – they never felt secure on their thrones. Competition among themselves drove them into building ever larger units and seeking ever greater shares in the market to achieve economies of scale. New entrants forced them to keep looking for oil so as not to be caught off guard. Rising nationalism among the producers constantly pressurized them to develop one source of supply from, say, Iran at the expense of another. What preserved the rule of the Seven Sisters was partly just historical accident. They held the all-important Middle East resources and the relentless growth in consumer demand required a level of investment in facilities with which they were able to keep pace out of their own earnings. In addition the persistence of surplus in supply always made the producers compete and the consumers reasonably relaxed.

The domination of so crucial an industry by so small a group of seemingly all-powerful companies did not, of course, go unchallenged. So great was the hatred of Rockefeller and his Standard Trust that, in Texas and the southwest, it was considered a betrayal of the flag to deal with them at all. The image of the small man fighting the big corporation, of the lone pioneer struggling to preserve his integrity and independence from the greed and deviousness of the majors became part of the folklore of the West. This was despite the fact that most of the independents who did become rich – the Hunts, the Gettys, the Hammers and the Pickenses of today – made their money more by dealing in other people's finds than by actually drilling themselves. An oil find like the East Texas field could literally make hundreds of millions, billions in the case of H. L. Hunt and John Paul Getty, for the owner of the lease. Even a small percentage of one of the Middle East concessions, whether it was the 5 per cent that Calouste Gulbenkian gained from the Red Line Agreement in Iraq or the concessions that Nelson Bunker Hunt and Dr Armand Hammer obtained in Libya could yield as much. The image created by the television series *Dallas* of wheeler-dealing, family feuds, corruption and lease swapping has been based on fact. Of the ten richest families in the United States, half have made their money out of oil.

Internationally, the American struggle of independent against major was translated into nation versus international company. Just as Rockefeller's control of the pipelines, rail cars and refineries of the United States infuriated the oil producers who found themselves at his mercy, so the oil-producing nations could never quite accept that their wealth should be so much under the control of the international companies who held the markets. Individual countries might have different ambitions at different times. The radicals, like Mexico in the thirties and forties, Algeria in the sixties and Libya in the seventies, wanted national control of all phases of the business at any cost. Iran under the Shah wanted off-take and high prices and if the oil companies seemed most likely to provide that, then so be it. Saudi Arabia preferred a policy of 'participation', a partnership between oil companies and the nation that would provide the country with the technology and the marketing arrangements under its own control. In the industrial countries like Britain and Norway the argument was complicated by the associated question of whether nationalism necessarily meant state companies.

Added to this continuous debate was the edge of human feeling. Tony Benn, twice Britain's Energy Secretary, remembers international companies descending on his office 'like emperors meeting a small local councillor'. Jim Akins, who worked in the US State

Department through the key events of the sixties and seventies, recalls that: 'When the oil companies came into the Middle East, and indeed for quite a number of years after, they considered that they were sovereign and that they dealt with the local leaders as sovereign equals. In one country where I served, the president of a major oil company came in to see the local ruler. The local ruler was pressing him for increased production because he needed more money and the oilman said, "Well, we'll do what we can for you. But, you know, it's very difficult."

'The ruler said, "But we're a friend of the company. We've done everything for the company. We're extremely friendly."

'And the oil company man said, "Why, of course, you're friendly. Everybody is friendly to us. The question is, how good a friend are you?" . . . and the rulers didn't forget this. When the situation changed, they remembered these slights and these humiliations ... and these linger and they dominate the thinking of the Oil Ministers in the OPEC countries today because some of them have been around for a long time.'

The changed circumstances came with a bang in the first oil shock, or the energy crisis as it became known, in 1973. Even without a squeeze on supply, it was doubtful whether the oil trade could go on without greater national control of its development. Sheikh Yamani's constant warnings during the sixties and early seventies that some form of partnership would have to be worked out if the oil companies were not to face outright nationalization were real enough. The 1973 crisis, and the dramas of the price negotiations with Libya and the Gulf countries in the preceding years, swung the balance of power round full circle. Suddenly it was the producers who were in the saddle. Local Scottish councillors could and did, treat the heads of visiting international oil companies as if they were petitioners, which they were and the Oil Ministers of OPEC could ask the oilmen just how good a friend *they* were.

The energy crisis didn't just represent a shift in the balance of power. To many it marked the end of an era and the dawn of a new age. Oil came to symbolize everything from the exploitation of the poor by the rich countries to the misuse of finite resources by the industrialized nations, the pursuit of profit at the expense of the unprotected and the power of the multinationals over technology and raw materials. Sheikh Yamani appealed for a new order in which the price of this essential commodity would be settled between producer and consumer and then ordered according to a formula which reflected inflation and currency values. Algeria became a spokesman for those who believed that control of oil should now become the axis around which the resource-rich countries could develop a new relationship with the manufacturing nations by forcing the latter to pay more highly for it.

Venezuela argued that oil should now be used to open up the North–South dialogue. Egypt, Kuwait and Iraq talked of oil as the political weapon that would impel the Europeans and the United States to bring Israel to a settlement of the Palestinian problem. The Shah of Iran referred to this 'noble product' as one that should be priced in relationship only to its finest uses in transport and petrochemicals. The Norwegians discussed the possibilities of oil becoming a bridge between the developed nations and the Third World, the Scots talked of breaking away from the UK on the back of oil, while the Shetlanders dreamt of becoming as rich as the Sheikhs of Araby.

Nor was this feeling that a seminal point in world economic development had been reached confined to the oil-producing countries. The environmentalists of the consuming nations now seized on oil as visible proof of a world that was depleting its resources and polluting its environment at an intolerable rate. The 'Club of Rome', a group of international think tanks founded after a meeting in Rome in 1968, commissioned a team of different specialists at the Massachussetts Institute of Technology to produce a model forecasting the ultimate effect of rising population and consumption on the world's resources of food, raw materials and energy. The conclusions, published in 1972 and 1973,[1] argued that growth in the world's economies would be ultimately limited by the lack of non-renewable resources, food as much as fuel, and that free-market economics could not provide the answers. The politics of scarcity became a modish subject as the newly fashionable trade of futurology predicted fundamental shifts in the world's power structure away from the resource-consuming to the resource-providing countries. Oil, it was said, was merely the harbinger of what was happening or going to happen with a wide range of commodities. The United States, Europe and, most especially, Japan were doomed to decline. The places of the future, as the international banks were not slow to reflect in their own lending, were the resource-rich, high-population countries like Brazil, Mexico, Venezuela, Iran and Indonesia. And when the more sceptical doubted these prophecies of revolutionary change, the second oil shock of 1979–80 came along to confirm them.

The second oil shock, as wiser heads such as Sheikh Yamani argued at the time, also ensured that the forces to solve such scarcity were set sharply in motion. If the seventies saw a sudden acceleration of long-voiced views of eventual supply constraints by industry experts such as David Barran of Shell and Harry Warman, BP's chief geologist, the eighties saw the growth of the developments that would bring supply and demand back into surplus. Demand fell not just for a year or two but consistently over five long years. Production from non-OPEC sources soared. Mexico and the North Sea doubled their exports. Oil was found in offshore India, as well as in Colombia, Malaysia and

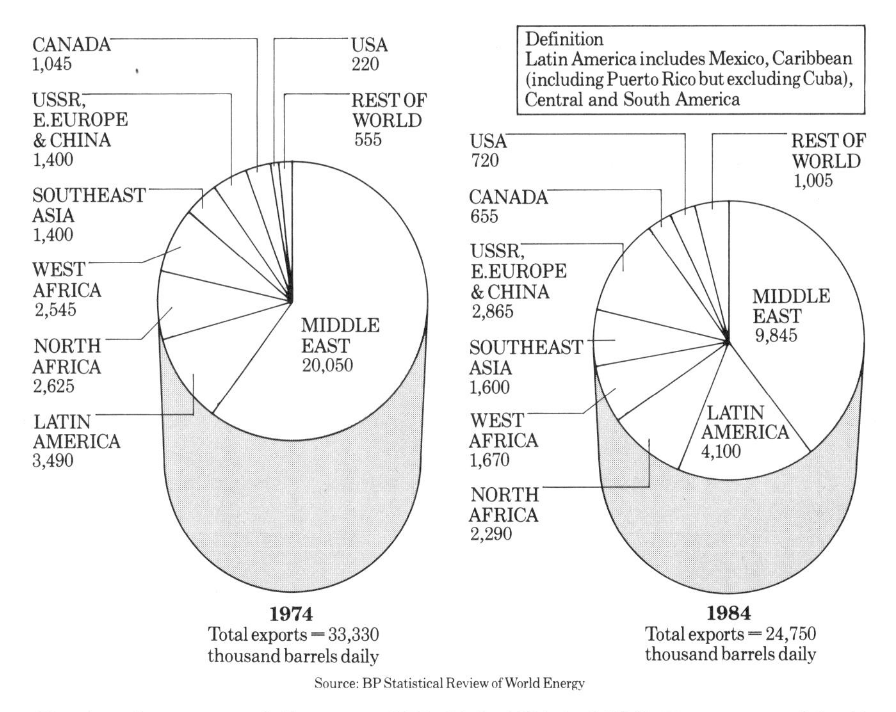

The changing pattern of oil exports, 1974–84. In 1974 the Middle East accounted for 60 per cent of all oil going into world trade. By 1984 its share had dropped to 40 per cent of a much-reduced total.

Egypt. Those with any sniff of oil rushed to develop it. Those without oil, particularly the poorer countries like Pakistan and Kenya, looked to everything from wood, gas and synthetic fuels from alcohol and garbage to solve their energy problems.

When OPEC met in London in March 1983 it was to face the harsh fact that the organization, having ridden the crest of shortages, was now taking the brunt of surpluses. OPEC output, which had touched 32 million barrels per day at the height of the 1973–4 crisis, was now down to 18 million barrels a day and dropping. Its only hope for holding prices was to institute production quotas which would convince the market that oil exports would be brought back into line with demand. When Saudi Arabia finally told its colleagues at an OPEC meeting in June 1985 that it was no longer willing to go on cutting output, it was no more than an acceptance that the policy of co-operative restraint wasn't working. Virtually everyone was cheating except Saudi Arabia, whose production had gone down from a peak 11 million barrels per day in 1979 to barely more than 2 million barrels per day in early 1985. Even a country as rich as Saudi Arabia couldn't absorb that kind of strain.

As the price plunged from $30 per barrel to $20 and then down as low as $10 in the spring of 1986, 1973 began to look less and less like a revolution and more like an aberration. From the perspective of 1986, it undoubtedly was. The wrench of the quadrupling of prices in the first energy shock, and even more their redoubling in the second oil shock, had given too many people the impression that oil was somehow set apart from any other commodity; that it was a resource that would defy all the ordinary laws of economics: and that the oil-rich would be able to live by their own rules on a different plane from the rest of the world.

The one devasting lesson of the post-crisis era was the foolishness of such reasoning and the folly of those who complacently believed that small countries could become great, that developing countries could become industrialized and that power could be reversed all at the touch of an OPEC wand. They forgot that oil's importance stems from the fact that it is an intrinsic part of the economy of consuming nations, not separate from it and that it is only one of the fuels in the energy balance. Oil was simply part of the inflationary crisis that built up in the world during the seventies, not the only part, nor even the main cause. When the price doubled again, however, it brought on a recession and government action in the West to cut back on inflation which reduced economic activity in general and hence the demand for oil. It also lessened the demand for traded goods of all kinds, making it that much more difficult for oil-producing countries to diversify out of oil into industrial goods. In retrospect, and as some commentators realized at the time, the fall of the Shah of Iran need not have given rise to the oil crisis that it did. There were adequate stocks and more than enough oil-production potential from other countries to cover the loss of Iranian exports. But, in a time of radical change in the structure of the industry and with the shift in ownership of oil from the companies to the producers, the customers panicked, the oil companies rushed into the fray to ensure supplies and the panic developed into a major crisis. The crisis, in turn, accelerated all the processes of adjustment that had already been set in motion by the first oil shock: the investment in alternatives to oil, the growth of energy conservation, the move of the developed economies from energy-intensive heavy industry to fuel-efficient electronics and service industries. In addition, such economic growth as there was, was being achieved with much less energy than before – the 'decoupling of growth and energy' as it was termed. Where once every 1 per cent growth in economic output brought with it nearly a 1 per cent growth in energy consumption, now the ratio dropped to 0·6–0·7 per cent consumption and less.

The figures told the story. Between 1965 and 1973 world consumption of primary energy (oil, gas, coal, nuclear power and hydroelectricity) went up by half from just under 4 billion tonnes of oil

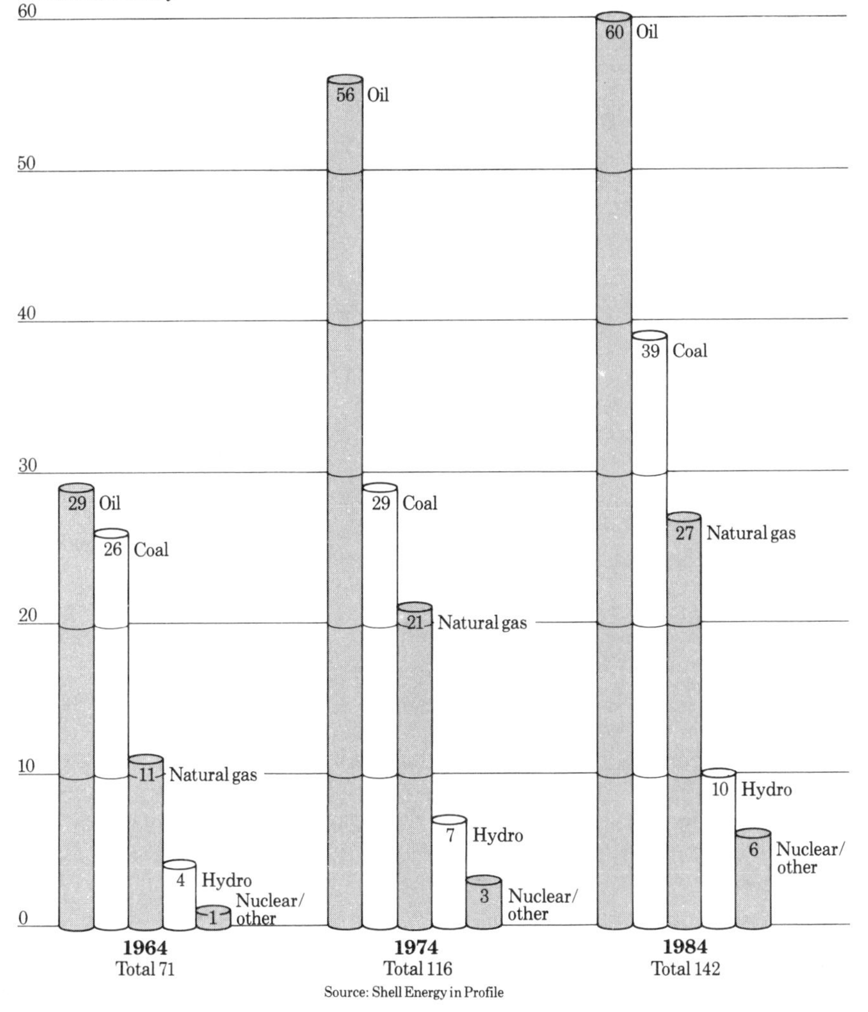

The energy balance over a generation. In 1964 oil's share of all world primary fuels was 40 per cent. Ten years later its volume had nearly doubled. By 1984 it had almost ceased to grow.

equivalent to just under 6 billion. Within that figure oil nearly doubled from 1·5 billion tonnes, or 38 per cent of the total, to 2·8 billion tonnes, or nearly 50 per cent. In the immediate aftermath of the first oil shock, primary energy demand went flat for a couple of years and oil consumption fell slightly. By the time of the second oil shock, total world energy demand had gone up by nearly 1 billion tonnes of oil equivalent to just under 7 billion tonnes, and oil took up 40 per cent of that increase. In the early eighties, total use of primary energy fell back again, only returning to its 1979 levels in 1984. But oil consumption was still down 10 per cent on its pre-crisis levels in the world as a whole and it had been reduced by around 20 per cent in Western Europe, the United States and Japan. Against the predictions by companies before the crises of what oil demand should have been by the mid-eighties, the actual out-turn was 50–60 per cent below forecast. The wonder in this situation to many was not that oil prices fell but that they took so long to do so.

Nor was there any sign of much relief on the horizon. So much of the reduction in energy use was associated with structural changes, from heavy to light industries, from bigger to smaller cars, from oil-fired stations to nuclear power, that few experts foresaw consumption ever returning to the kind of annual rates of increase seen in the seventies, even if there was an upsurge in economic growth.

Without any prospect of a dramatic growth in demand and with oil prices falling, a shake-out of the market and of the industry was inevitable. The kind of huge-scale investment in heavy-cost oil sources – from oil shale, tar sands and coal – was the first to go. The only way this could ever have been justified was if oil demand had continued to produce almost impossible shortages. The next thing to happen was a slimming down in the industry itself. Many of the smaller companies, which had grown fat on the sudden rise in prices in the United States and the North Sea, had to shut down or merge. So did some of the big companies. The assault on the majors by T. Boone Pickens, using sophisticated take-over tactics, was based on the simple observation that, at a time of falling prices, companies were failing either to replace their crude oil at a low enough cost or to diversify into other industries effectively. By the end of 1985 one of the Seven Sisters had gone, when Gulf Oil was merged with Standard Oil of California, and the future of at least two of the others was in doubt.

With the fall in prices, the higher cost producing areas also came under pressure. The 'stripper wells' of the United States (wells capable of producing only a few barrels per day) started to shut down in their thousands. The number of drilling rigs in the country was halved and Texas and Oklahoma faced severe economic problems. The capital expenditure budgets of the bigger companies in 1986 was reduced by an average 20–30 per cent and in some cases halved. New investment in the North Sea, in Alaska and the frontier area of the Arctic and the Atlantic was reviewed and, in many cases, halted entirely.

The beneficiaries of this seemed likely to be – as they always are in a period of shake-out – the stronger companies and the lower cost producers. Exxon, Shell and, to a slightly lesser extent, BP looked healthy. Saudi Arabia, Kuwait, Iraq and Abu Dhabi appeared to be the fittest of the producers. The oil picture, ironically, is now moving back in time to the sixties, before the North Sea and before even the rise of Libya and the independents.

That is the problem. For, while the first oil shock may have been an aberration in some senses, at least in the ambitions that it set off, it also sounded a clear warning signal. If demand had gone on rising as it had before the crisis, there would have been a real shortage instead of one artificially induced by producer cutbacks. Saudi Arabia, with its limited internal needs, would have only overheated its economy by going on expanding, as would Kuwait. The North Sea and other areas could not have been developed without higher prices. The world would have faced, as it did to an extent, intolerable pressures of inflation, pollution and resource scarcity if it had continued to grow without structural adjustment.

To that extent 1973 was a corrective to the excessive expansion pressures building up in the sixties and early seventies, just as 1986 may be a necessary corrective to similarly excessive pressures and expectations built up by the two successive oil shocks. It may even be beneficial for some of the producing countries. This is the opinion of Prince Abdullah Bin Feisal Bin Turki, one of the younger members of the Saudi royal family. After 1973–4, he suggests, 'We went through a period of huge boom for about eight years or so, when there was a lot to be done and done very quickly. We had a huge amount of income compared to what we had. But we had a huge country, as big as Western Europe but with only about 7 million people in it. So we went through a lot of economic changes, a lot of development changes and undoubtedly a lot of social change. I cannot think of anywhere in the world where this degree of change could be compared.'

The danger is, that in correcting the excesses of 1979, people are forgetting the lessons of 1973. Despite all the drilling of the last decade, the dominance of Middle East reserves in the world's oil picture remains virtually unaltered. One of the clearest warning lights before the two energy crises was the world's failure to find anything like the number of giant fields it needed to replace the oil being taken out of the ground. That, as the Chinese example shows, remains largely true today. Increased production from outside the Middle East has come not from new discoveries so much as from high-cost fields already known and made viable by high prices. Remove those high-cost sources and you are back to dependence on the Middle East and, according to Sheikh Yamani, to dependence on OPEC.

Many industrial experts agree with him. At lower prices, non-OPEC and some non-Middle East OPEC sources will decline, demand will revive and, at some point, the world will be back to a supply crunch. The lower the price, the sooner it will come.

It is not an analysis that everyone favours, however. A number of commentators, like Professor Peter Odell of Erasmus University, Rotterdam, feel that: '. . . oil is not scarce. The price is falling and now the options for the future in terms of which way the industry will go is no longer a matter of OPEC control nor of control by the Seven Sisters, but a matter of how we organise the system in the context of a very large number of actors in this emerging industry.' Professor Odell, like Sheikh Yamani, wants to see a better relationship between OPEC and the industrialized countries to 'make sure that our system doesn't collapse'. His fear is of a price plunge that would upset the world's banking system and cause political problems in a number of key countries. Others argue that this is wrong, that efforts to produce stability in the market have failed and that it is only through letting the market find its own level that a long-term price and role for oil can be established. The instability of oil, they argue, is because of government intervention in the past, whether by the OPEC cartel or by the American decision to control imports and prices, not for lack of it.

The difficulty of this argument is that oil is, at bottom a commodity. It has to obey the laws of supply and demand like any other. But it has taken on a role in world trade quite unlike any other. The nature of oil consumption makes it essential for economic growth. As yet there are no substitutes for oil as a transport fuel. Nor is there any commodity which has such a profound effect on a country's balance of payments, its exchange rates and its reserves. The economics of oil have always made it peculiar. At one end of the cost scale is the Middle East, which can produce at $1 per barrel or less but has a limit to its self-interest in pouring out as much oil as the world might need. At the other end are the new finds in the North Sea and the Arctic which can cost as much as $20 to $30 per barrel but have been found in countries which, for strategic reasons, may wish to keep on producing oil even if it is uneconomic. The nature of its discovery makes it subject to dramatic lurches in price – and always has.

Given these pulls on oil it is hardly surprising that it has become as much a political as a commercial commodity. OPEC still has potential control on price and supply should it wish to exercise it. The organization accounts for nearly two-thirds of the world's oil exports, if only 40 per cent now of its production. The collapse of prices during 1986 was brought about by Saudi Arabia's decision to raise production, not by the iron laws of the market. It could still be reversed by an OPEC decision to lower output to the point where prices have to strengthen.

Oil is also political because the consumers have come to regard it as too vital a commodity not to have influence in some way or another. Those with energy resources, like the United States and Britain, wish to sustain them in the event of a crisis. Those without, like Japan, wish to reduce their vulnerability, even at a cost. One of the principal reasons why the world did lurch to two successive oil crises in the seventies was the dramatic move by the United States into the import market for oil. One of the reasons that some fear that oil may return to a crisis in the nineties is the decline in Russian oil production, which peaked for the first time in 1984–5 and is likely to fall sharply over the next decade. The strategic issue of oil-import dependence hasn't gone away since the energy crisis.

Nor has the environmental question. The Chernobyl nuclear disaster in Russia in April 1986, the worst nuclear accident in the industry's history, proved a devastating and tragic reminder of the penalties of an overhasty move away from traditional fuels. Yet the consequences of continuing to burn up larger and larger volumes of fossil fuels each year are also potentially hazardous. Research since 1973 suggests that some of the fears of oil spills on the marine environment and oil explosions have been exaggerated. Concern about the impact of growing carbon dioxide emissions, together with those of other chemicals, on the earth's atmosphere have, however, tended to become more real with time. Evidence collected by the United Nations and other bodies now points to the definite prospect that the 'greenhouse effect' of these emissions – the creation of a layer that lets in heat but doesn't allow it out easily – will raise the temperature of the world by several degrees and the level of the sea by as much as a metre by the end of the century.

In the physical environment, as in the economic environment, oil has to find a balance. It is part and parcel of the processes of industrialization and growth which produced such strains in 1973 and could still do so again. Yet balance and perspective are the two features that have been singularly lacking in oil's history. It has always been a boom-and-bust business, ever since the days of Colonel Drake. It is still true today when the industrialized world has come to rely on it, nations have become rich on it and individuals and companies have tried to corner it.

For those who believe that market forces will bring the best solution if only left free to work their way, oil has been a classic example of the follies of market manipulation and the foolishness of forecasts. For those who believe, in contrast, that the root of all oil volatility has been in mankind's eternal desire to buy more as soon as it is cheap enough, the present situation can only serve to repeat the vicious cycle of feast and famine that has dogged oil's history from the start.

The one point that has not changed is that those without oil still want it and those with oil are never quite content with it. In all the discussion about the Middle East, OPEC, and the North Sea, it is often forgotten that the two largest oil producers in the world are the two superpowers – Russia and the United States. They were the first major oil producers in the world and they remain by far the largest still, each producing about 20 per cent of the world's total output at over 10 million barrels per day, yet both fearful of declining production levels and reserves after so long a history of intense exploitation. Russia is pushing to the east, driving its gas and searching now in the deep waters of the Barents Sea in near Arctic conditions. Its fear is always that oil, its largest foreign-currency earner, may decline too rapidly to allow gas or nuclear power to take up the momentum. In a centrally planned economy, the mistakes of government bureaucracy can be all the greater. In the United States, on the other hand, the fear is that the market itself may move too quickly and undermine the oil eked out of old fields and that their country, which learnt so bitter a lesson from its import dependence in 1973 and 1979, may be forced to return to those days unless Government acts to protect its industry from imports and its non-oil investments, such as research into synthetics, from liquidation.

Yet over in the Pacific is China, the most populous country in the world and edging towards the economic leap that would make it the next superpower in the coming century. By then it could have 1 billion customers and an oil thirst potentially as great, if not greater, than the United States or Russia. Yet its present oil production of a little over 2 million barrels per day is adequate neither to meet the pent-up demand for transport and power, nor to provide the export earnings that would help kick-start the economy. China needs the oil boost, while the oil industry, and the major oil companies, need China to get that boost to give them the kind of fillip they need in a world of stagnant demand. In the hotel suites of Beijing and the offices of the central and provincial ministries negotiations take place, as they always have, between oil companies anxious to get a share of the action but eager to ensure the profits that will reward this risk and fund the next one elsewhere, and the national Government, wanting to gain the enterprise and the trading skills of the foreign companies but desperately anxious not to lose control of their most precious potential resource. It is a match that has been played in a hundred countries and will go on being played as long as oil is a gamble as well as a business. For the country and for the company that finds it, the wealth can be limitless and the problems countless. China with oil has the chance to accelerate into the twenty-first century. Without it, the country must simply plod on its way.

That remains the difference between oil and other commodities. Only oil has the power to change the balance of power.

Notes

Many of the extracts in the book have been taken from interviews conducted for the television series by the producers Ted Brocklebank and Bjorn Nilsen. Other sources are given below.

Chapter 1: 'God Bless Standard Oil'

1. For a comprehensive summary of the first beginnings and early technology of oil, see Robert O. Anderson, *Fundamentals of the Petroleum Industry* (Weidenfeld & Nicolson, 1984).
2. Quoted in H. Dolson, *The Great Oildorado* (Random House, 1959).
3. A barrel measurement was based on the wooden Pennsylvania barrels used to transport it. It is equivalent to 42 US gallons and 35 imperial gallons.
4. Quoted in Richard O'Connor's vivacious account of the pioneers of oil, *The Oil Barons* (Hart-Davis MacGibbon, 1972).
5. Quoted in Allan Nevins, *Study in Power: John D. Rockefeller, Industrialist and Philanthropist* (Charles Scribner & Sons, 1940–53).
6. *ibid.*
7. Quoted in O'Connor.
8. Quoted in Christopher Tugendhat's historical chapters in *Oil: The Biggest Business* (Eyre & Spottiswoode, 1968) which gives a clear and balanced account of Rockefeller.
9. Quoted in Anthony Sampson, *The Seven Sisters* (Hodder & Stoughton, 1975).
10. Quoted in Ruth Sheldon Knowles, *The Greatest Gamblers: The Epic of American Oil Exploration* (McGraw-Hill, 1959).
11. Quoted in O'Connor.
12. *ibid.*

Chapter 2: Floating to Victory

1. Quoted in Robert Henriques, *Marcus Samuel, First Viscount Bearsted and Founder of Shell* (Barrie and Rockliff, 1960).
2. Henri Deterding with Stanley Naylor, *An International Oilman* (Harper, 1934).
3. For a detailed and clearly written summary of British policy on oil in the early years, see Geoffrey Jones, *The State and the Emergence of the British Oil Industry* (Macmillan, 1981).
4. Quoted in Jones.
5. *ibid.*
6. *ibid.*
7. Nubar Gulbenkian, *Pantaraxia: An Autobiography of Nubar Gulbenkian* (Hutchinson, 1965).

8. Quoted in Richard O'Connor, *The Oil Barons* (Hart-Davis MacGibbon, 1972).
9. Quoted in Jones.
10. Quoted in Anthony Sampson, *The Seven Sisters* (Hodder & Stoughton, 1975).
11. Quoted in Christopher Tugendhat, *Oil: the Biggest Business* (Eyre & Spottiswoode, 1968).
12. Quoted in Jones.
13. Quoted in O'Connor.
14. Quoted in Jones.

Chapter 3: Sisters Under Siege

1. Quoted in Richard O'Connor, *The Oil Barons* (Hart-Davis MacGibbon, 1972).
2. The story is told in John Brooks, *The Games Players* (Times Books, New York, 1980).
3. Quoted in John O. King, *Joseph Stephen Cullinan* (Vanderbilt, 1970).
4. See M. Adelman, *The World Petroleum Market* (Johns Hopkins University Press; 1972).
5. Quoted in Leonard Mosley, *Power Play: Oil in the Middle East* (Weidenfeld & Nicolson, 1973).
6. Quoted in Stephen Hemsley Longrigg, *Oil in the Middle East* (Oxford University Press, 1954).
7. See interviews in Brian Lapping, *End of Empire* (Granada Publishing, 1985).
8. Quoted in Mosley.

Chapter 4: 'Charter for Change'

1. Quoted in Robert O. Anderson, *Fundamentals of the Petroleum Industry* (Weidenfeld & Nicolson, 1984).
2. Quoted in George W. Stocking, *Middle East Oil: A Study in Political and Economic Controversy* (Allen Lane, 1971).
3. See R. W. Ferrier, *The History of the British Petroleum Company:* vol. 1 *The Developing Years 1901–32* (Cambridge University Press, 1982).
4. Quoted in Stocking.
5. H. St John Philby, *Arabian Jubilee* (Robert Hale, 1952).
6. The details are in Philby's *Arabian Jubilee* and *Arabian Oil Ventures* (Middle East Institute, 1964).
7. Quoted in Stocking.
8. See Leonard Mosely, *Power Play: Oil in the Middle East* (Weidenfeld & Nicolson, 1973).
9. For an account of the meeting and negotiations with Gulbenkian, see *Pantaraxia: An Autobiography of Nubar Gulbenkian* (Hutchinson, 1965).
10. Quoted in Mosely.
11. Taken from an interview with the Middle East Economic Survey and quoted in Ian Seymour, *O.P.E.C.: Instrument of Change* (Macmillan, 1980).
12. *ibid.*
13. Quoted in Christopher Tugendhat, *Oil: the Biggest Business* (Eyre & Spottiswoode, 1968).

Chapter 5: The Independents

1. The *Observer*, 6 April 1986.
2. Quoted in Stephen Fay, *The Great Silver Bubble* (Hodder & Stoughton, 1982).
3. The details are in *The Great Silver Bubble* and in Harry Hurt's *Texas Rich* (Orbis, 1981).
4. Quoted in Hurt.
5. *ibid.*
6. J. Paul Getty, *My Life and Fortunes* (Allen & Unwin, 1963).
7. Quoted in Robert Lenzner, *Getty, the Richest Man in the World* (Hutchinson, 1985).

Chapter 6: 'The Devil Gave Us Oil'

1. Quoted in Richard O'Connor, *The Oil Barons* (Hart-Davis MacGibbon, 1972).
2. Frank C. Hanighen, *The Secret War* (Routledge, Kegan Paul, 1934) quoted in O'Connor.
3. For the details see Geoffrey Jones, *The State and the Emergence of the British Oil Industry* (Macmillan, 1981).
4. Adrian Hamilton in the *Observer*, 13 May 1984.
5. Interview with the author.

Chapter 7: A Place in the Sun

1. Paper presented to the *Financial Times*' North Sea Conference, London, September 1972.
2. Quoted in Clive Callow, *Power from the Sea: Search for North Sea Oil and Gas* (Gollancz, 1973).
3. The full details are set out in an article by Jonathan Wills in the *Shetland Times*, 3 January 1986.
4. William Keegan, *Britain Without Oil* (Penguin, 1985).

Chapter 8: The Global Gamble

1. See Dennis L. Meadows, *The Limits to Growth* (Earth Island, 1972) and *The Dynamics of Growth in a Finite World* (Wright-Allen Press, 1973).

Notes for Further Reading

A general account of the major oil companies and the crises of the sixties and seventies – and a source for much of the series – can be read in Anthony Sampson's *The Seven Sisters* (Hodder and Stoughton, 1975). A lucid and balanced account of the history of oil is contained in Christopher Tugendhat's *Oil: the Biggest Business* (Eyre & Spottiswoode, 1968). General explanations of the industry, together with the broad background are found in Robert O. Anderson's *Fundamentals of the Petroleum Industry* (Weidenfeld & Nicolson, 1984), Shell's *The Petroleum Handbook* (6th edition Elsevier, 1983) and BP's *Our Industry Petroleum* (BP, 1977).

For the early history of oil in the United States, Allan Nevins' *Study in Power: John D. Rockefeller, Industrialist and Philanthropist*, 2 vols (Charles Scribner & Sons, 1940–53) remains a classic study of the oil giant, while Ida M. Tarbell's *The History of the Standard Oil Company* (McClure's, 1904) remains the classic criticism. Richard O'Connor's *The Oil Barons* (Hart-Davis MacGibbon, 1972) gives a colourful description of many of the pioneers, as does Ruth Sheldon Knowles' *The Greatest Gamblers: The Epic of American Oil Exploration* (McGraw-Hill, 1959) and James A. Clark and Michael T. Halbouty's *Spindletop* (Random House, 1972).

For the early history of international oil, see Robert Henriques' sympathetic account in *Marcus Samuel, First Viscount Bearsted and Founder of Shell* (Barrie and Rockliff, 1960). Gulbenkian is treated in both his son, Nubar Gulbenkian's *Pantaraxia: An Autobiography of Nubar Gulbenkian* (Hutchinson, 1965) and Ralph Hewins' *Mr Five Per Cent: the Biography of Calouste Gulbenkian* (Hutchinson, 1957). Geoffrey Jones' *The State and the Emergence of the British Oil Industry* (Macmillan, 1981); T.A.B. Corley's *A History of the Burmah Oil Company* (Heinemann, 1983) and R. W. Ferrier's *The History of the British Petroleum Company*, vol. 1, The Developing Years 1901–1932 (Cambridge University Press, 1982) have all revealed much new information on the first decades of British involvement in the industry, The standard work on Mattei remains Paul Frankel's *Mattei: Oil and Power Politics* (Faber & Faber, 1966).

The history of Middle East oil finds an entertaining narrative in Leonard Mosley, *Power Play: Oil in the Middle East* (Weidenfeld & Nicolson, 1973). Stephen Hemsley Longrigg gives an ex-Anglo-Iranian viewpoint in *Oil in the Middle East* (Oxford University Press, 1954) while George W. Stocking *Middle East Oil: A Study in Political and Economic Controversy* (Vanderbilt University Press, 1970 and Allen Lane, 1971) gives more of an American viewpoint. Mossadeq's Iran gets an excellent chapter in Brian Lapping's *End of Empire* (Granada Publishing, 1985), while the weaknesses of the Shah's Iran are percipiently detailed in Robert Graham's *Iran: Illusion of Power* (Croom Helm, 1979). The Shah's fall and the Ayatollah's rise are described in Anthony Parsons' *The Pride and the Fall* (Jonathan Cape, 1984) and Shaul Bakhash's *The Reign of the Ayatollahs* (Basic Books Inc, 1985). Libya gets scholarly treatment in Frank C. Waddams' *The Libyan Oil Industry* (Croom Helm, 1980) and a broader survey in John K. Cooley's *Libyan*

Sandstorm: Complete Account of Qadhafi's Revolution (Sidgwick & Jackson, 1983).

Saudi Arabia has provided a number of accounts on oil, including David Holden and Richard Johns' *The House of Saud* (Sidgwick and Jackson, 1981), Robert Lacey's *The Kingdom* (Fontana, 1982), and St John Philby's *Arabian Jubilee* (Robert Hale, 1962) and *Arabian Oil Ventures* (Middle East Institute, 1964). He has also received full biographical treatment in Elizabeth Monroe's *Philby of Arabia* (Faber & Faber, 1973).

More recent history is dealt with in Ian Seymour's *O.P.E.C.: Instrument of Change* (Macmillan, 1980); Yusif A. Sayigh, *Arab Oil Policies in the 1970s* (Croom Helm, 1983); James M. Griffin and David J. Teece (ed); *Organization of Petroleum Exporting Countries Behaviour and World Oil Prices* (Allen and Unwin, 1982). The issues raised by the 1973 crisis are discussed, from various viewpoints, by Jack Anderson in *Oil* (Sidgwick and Jackson, 1984), a conspiracy view of the history; John M. Blair, *The Control of Oil* (Random House, 1976), a trenchant criticism of the oil company oligopoly by a former economist with the Federal Trade Commission; and *OPEC: Twenty Years and Beyond*, a symposium on the organization's future, edited by Ragael El Mallakh (Westview Press Inc., 1982). *Energy Policy in Perspective: Today's Problems, Yesterday's Solution*, ed. D. Craufurd Goodwin (The Brookings Institution, 1981), gives a concise summary of the policies followed by the various consumer governments in response to the oil crises.

Among the studies of the American independents, Harry Hurt's *Texas Rich* (Orbis, 1981), a study of the Hunts of Dallas, stands out for its research and readability. The Hunts' silver scam is covered in detail in Stephen Fay's *The Great Silver Bubble* (Hodder & Stoughton, 1982). Getty has had two recent biographers: Robert Lenzner, *Getty, the Richest Man in the World* (Hutchinson, 1985) and Russell Miller, *The House of Getty* (Michael Joseph, 1985).

The North Sea has been the subject of a number of books including: Bryan Cooper and T. F. Gaskell, *North Sea Oil – The Great Gamble* (Heinemann, 1966); Clive Callow, *Power from the Sea: Search for North Sea Oil and Gas* (Gollancz, 1973), James R. Nicolson, *Shetland and Oil* (William Luscombe, 1975); Adrian Hamilton, *North Sea Impact, Off-Shore Oil and the British Economy* (International Institute for Economic Research, 1978); D. I. MacKay and G. A. Mackay, *The Political Economy of the North Sea* (Martin Robertson, 1975); Aubrey Jones, *Oil: The Missed Opportunity* (André Deutsch, 1981); Guy Arnold, *Britain's Oil* (Hamish Hamilton, 1978), G. Corti and F. Frazer, *The Nation's Oil: Story of Control* (Graham & Trotman, 1983) and William Keegan, *Britain Without Oil* (Penguin, 1985).

Oil continues to provide a constant flow of books and the proceedings of conferences on its future and on its arguments. For those interested in forecasts, old and new, Peter Odell and Kenneth Rosing have published a study of their model on resources, *The Future of Oil: A Simulation Study of World Supply and Demand 1980–2080* (Kogan Page, 1980); Robert Stobaugh and Daniel Yergin have edited the report of the Energy Project at the Harvard Business School, *Energy Future* (Random House, 1980); Daniel Yergin with Martin Hillenbrand is also author of *Global Insecurity, a Strategy for Energy and Economic Renewal* (Houghton Mifflin, 1982). The International Energy Agency has also published a series of reports and forecasts, including *World Energy Outlook* (Paris, 1982). Among conferences, Paul Tempest has edited the report of the 1980 Conference of the International Association of Energy Economists, *International Energy Options: An agenda for the 1980's* (Graham & Trotman, 1981).

INDEX

Compiled by Diana LeCore

Page numbers in *italic* refer to line drawings

ILLUSTRATION ACKNOWLEDGMENTS

Between pages 64 and 65 (black and white)

BBC Hulton Picture Library/The Bettman Archive: v (above); British Petroleum: i (above right), ii (above and below), iii (above, top right); Dennis Coutts: viii; Drake Well Museum: i (above left and below); ENI: iv (above right); Esso: iii (above, bottom right); Grampian Television: M. Czaky vii (below); The Hunt Family Archives (Hunt Oil Company): vi (above); Popperfoto: iv (below), v (below), vi (below); Shell International Petroleum Company: iii (above left and left); Frank Spooner Pictures: vii (above); Topham: vi (above right).

Between pages 128 and 129 (colour)
Mary Evans Picture Library: v (above and below); Grampian Television: M. Czaky iv (above) and viii (above), Jim Gibson iv (below), Rod Jordan ii–iii; Occidental International Oil Inc.: vii (above); The Rockefeller Archive Center: i; UK AEA: viii (below); Wharton Williams: vii (below); Zefa: vi.

Line drawings by Eugene Fleury

Permission to use the diagram on page 27 was kindly granted by the University of Oklahoma Press.

The extract from H. St John Philby's *Arabian Jubilee* on page 83 is quoted by kind permission of Robert Hale and David Higham Associates Ltd.